Nature of Research: Scientific
Authority of Research: Academic
Fields of Research: Physics and Telecommunications Engineering
Types of Research: Theoretical Analysis, Mathematical Formulation and Numerical Simulation

Title/Subtitle:

Reflection and Transmission of Electromagnetic Plane Waves on and through Isotropic and Anisotropic Uniaxial Plates

According to a Coordinate-Free Analysis Approach

First Edition - Third Release

By:
Ibrahim Ibrahim
December 2018

First Published in:
2006

ISBN-13:978-1791530709

Preface

This work mathematically[1] formulates the reflection and transmission coefficients of the *plane waves*[2] which propagate in free space through[3] two different kinds of *stationary media*[4]: isotropic and anisotropic uniaxial plates. This was achieved under the condition that the optic axis $\widehat{\mathbf{c}}$ is parallel to the plane of incidence.

In this thesis, RC_2 *pulse*[5] was chosen as a preferred signal with which the behavior of the coefficients was examined, and Fast Fourier Transform (FFT) tool from *Matlab*® was used to simulate the reflected and transmitted signals at the isotropic and uniaxial media interfaces in *time domain* after analytically modeling the system in *frequency domain*.

In contrast to almost all of the published literature on topics of Electromagnetism, this thesis applies a *coordinate-free analysis approach* based on (and completely relying on) Chen's work[6]; abandoning therewith the conventional analysis that depends on an already predefined coordinate system in advance.

---

[1] See Appendix C.1.
[2] See Appendix B
[3] See Appendix F
[4] See Appendix E
[5] See Appendix A
[6] [Chen (1983)]

Introduction

In the first chapter, the basic properties of the *monochromatic*[7] plane wave were presented alongside the laws of reflection and refraction which are derived from the *boundary conditions*[8]. At the end of this chapter I presented Chen's analytical model [Chen (1983)] and then used it in my own specific application in the second and third chapters (sections 2.3 and 3.8) to further apply the same analysis on the two cases of the isotropic and the anisotropic uniaxial plates respectively. After I presented Chen's system of equations (in chapters 2 and 3) and before I introduced my own application at the end of these chapters, I have literally derived step by step (and before the reader's eyes) all of Chen's equations to facilitate a much easier access into this material (and more specifically to rectify a major mistake in Chen's derivations) for both of the professionals and beginners of Science and Engineering alike. I have also computationally simulated the mathematically-dynamic model of my application (for both cases in the above-mentioned sections). Furthermore, I have visualized Chen's own mathematically-static abstract equations (in section 3.7.5) and corrected his aforementioned mistake (in equation 3.236).

Chapter 2, hence, handles the behavior of the plane wave in the isotropic medium and shows the reflection and transmission thereof at the plane of interface. Then the topic advances into the reflection and transmission of the plane wave through the isotropic plate. FFT tool were used to produce and simulate the final reflected and transmitted signals from the RC_2 test pulse which propagates satisfying normal incidence conditions through the plane of interface.

The behavior of the plane wave was then examined in crystals and were presented in Chapter 3 where a general look into the crystals is made until section 3.7. In this part of the thesis, Uniaxial properties are analyzed in depth starting with the dielectric tensor of the medium itself; the directions of field vectors and wave vectors; and ending with the reflection and transmission coefficients which were formulated based on different orientations of the optic axis.

---

[7] See Appendix D.
[8] See Appendix C.

I have formulated the transmission and reflection coefficients of the plane wave which penetrates the uniaxial plate for three different cases:
The first is when the optic axis is perpendicular to the normal of the interface.
The second is when the optic axis is parallel to the normal of the interface (where it will be shown that the latter has the same behavior as that of the isotropic-isotropic interface).
And finally the third case when the optic axis is tilted 45° away from the normal of the interface; the difference between both perpendicular and parallel permittivities here equals zero, and the plate behaves as if it were not having this tilt in its optic axis.

Contents

List of Figures

Part I.

PLANE WAVES

1

Monochromatic Plane Waves Properties

1.1. General Properties

A monochromatic plane wave field has a sinusoidal variation in space and time with an instantaneous vector $\underline{E}$:[1]

$$\underline{E} = \Re(\mathbf{E}e^{-i\omega t}) \tag{1.1}$$

where the complex vector **E** depends on the position vector **r** and is given by [2]

$$\mathbf{E} = \mathbf{E}_0 e^{i\mathbf{k}\cdot\mathbf{r}} \tag{1.2}$$

$\mathbf{E}_0$ is a complex-constant amplitude vector independent of **r**.

k is the wave vector=$k\widehat{\mathbf{k}}$, with k as the wave number and $\widehat{\mathbf{k}}$ is the wave normal.

$$k = \frac{w}{c}\sqrt{\mu\varepsilon} \tag{1.3}$$

where: ω is the angular frequency, c speed of light, μ permeability and ε is the permittivity.

Equation (1.1) has a constant phase of

$$e^{i\mathbf{k}\cdot\mathbf{r}}e^{-i\omega t}$$

[1] Some books use the convention $\mathbf{E}(\mathbf{R}, t)$ rather than the one which has been used here to distinguish the vector in time domain from that which lies in frequency domain.
[2] The vector in frequency domain is **E**, and in time domain is $\underline{E}$.

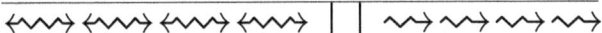

yielding to

$$\phi = \mathbf{k} \cdot \mathbf{r} - \omega t \tag{1.4}$$

In Figure 1.1 below, ϕ_1 is a plane which is perpendicular to the unit vector $\hat{\mathbf{k}}$ and is located away from the origin at a distance ζ. And in Figure 1.2 below we see the following relation

$$\begin{aligned} \cos\theta_1 &= \frac{p}{k_1} \\ \cos\theta_2 &= \frac{p}{k_2} \end{aligned}$$

Dividing the above equations over one another gives:

$$\begin{aligned} \frac{\cos\theta_1}{\cos\theta_2} &= \frac{k_2}{k_1} \\ \Longrightarrow \quad k_2\cos\theta_2 &= k_1\cos\theta_1 \end{aligned}$$

Hence,we have

$$\hat{\mathbf{k}} \cdot \mathbf{r}_1 = \zeta_1 = \frac{\phi_1 + \omega t_1}{k} \tag{1.5}$$

Where ϕ_1 is defined as the surface of constant phase which maintains its constant value when t or ζ increases. And in Figure 1.1, we see that

$\mathbf{r}_{11} \cdot \hat{\mathbf{k}} = \mathbf{r}_{12} \cdot \hat{\mathbf{k}} \quad \Longrightarrow \quad \zeta_{11} = \zeta_{12}$
but if ζ_1 changes to $\zeta_2 \quad \Longrightarrow \quad \zeta_{21} = \zeta_{22}$

delivering to us the second surface ϕ_2 of this *uniform plane wave*[3]. Equation (1.5) is then modified into:

$$\hat{\mathbf{k}} \cdot \mathbf{r}_2 = \zeta_2 = \frac{\phi_2 + \omega t_2}{k}$$

with the distance ζ_2 (for the second surface) away from the origin.

The distance ζ gives the constant phase, as it travels from one point to another and in the direction of $\hat{\mathbf{k}}$ with a a wave velocity called *phase velocity*:

$$\nu_p = \frac{d\zeta}{dt} = \frac{\phi + \omega t}{k} = \frac{\omega}{k} \tag{1.6}$$

where $\quad \frac{d}{dt}(\frac{\phi}{k}) = zero$

If the difference between two distances $(\zeta_2 - \zeta_1)k$ equals to 2π radians, then:

[3] See Appendix B.

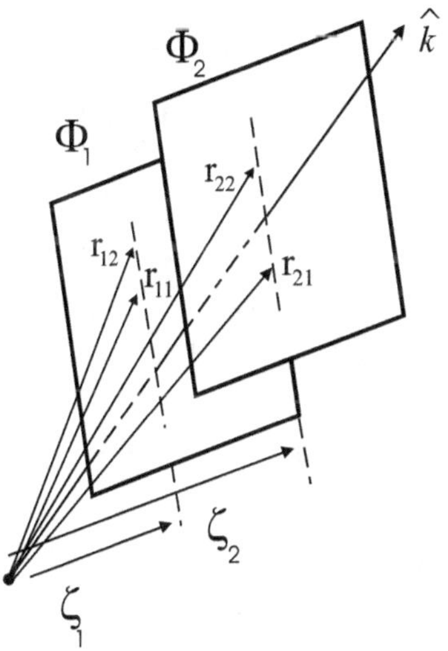

Figure 1.1.: Uniform Plane Wave Propagating in the Direction of $\widehat{\mathbf{k}}$.

$\Longrightarrow \quad (\phi_{2s} + \omega t - \phi_{1s} - \omega t) = \phi_{2s} - \phi_{1s}$

The s above in ϕ_{2s} *and* ϕ_{1s} stands for same amplitude of the wave on these planes of constant phase. This means that in order to reach the surface of ϕ_{2s} from ϕ_{1s}, either

1) an increase of ζ is needed without increasing t.

$$\phi_{2s} - \phi_{1s} = 2\pi = \mathbf{k} \cdot (\mathbf{r}_2 - \mathbf{r}_1) = k\lambda$$

2) an increase of t is needed without increasing ζ.

$$t_1 - t_2 = T = \frac{2\pi}{\omega} = \frac{1}{F} = \frac{1}{(f_1 - f_2)}$$

3) an increase of both t and ζ is to be maintained in the relation

$$2\pi = k\lambda + \omega T$$

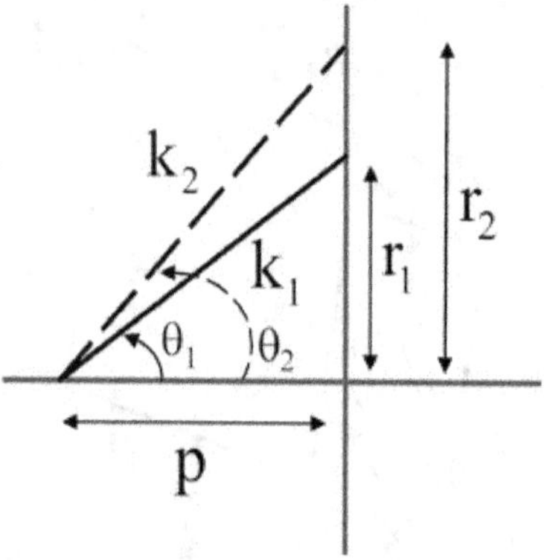

Figure 1.2.: The Angles.

1.2. Refractive Index

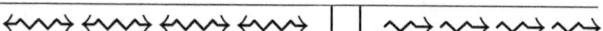

The term *refractive index vector* $\mathbf{n}$ is used sometimes and defined by

$$\mathbf{n} = \frac{c}{\omega}\,\mathbf{k} \tag{1.7}$$

with a magnitude of

$$n = \frac{ck}{\omega} = \sqrt{\mu\varepsilon} \tag{1.8}$$

The phase of the wave may be written as

$$\phi = \omega(\frac{1}{c}\,\mathbf{n}\cdot\mathbf{r} - t) \tag{1.9}$$

From section (2.2), we notice that the transmitted and incident wave vectors can also be expressed in the following relation:

$$\frac{sin\theta_i}{sin\theta_t} = \frac{k_t}{k_i} = \frac{n_t}{n_i} = n_{21} \tag{1.10}$$

where n_{21} is the *relative index of refraction* of the second medium with respect to the first. Equation (1.10) is known as Snell's law of refraction.

1.3. Electromagnetic Field Vectors

Complex Maxwell equations in frequency domain are:

$$\begin{aligned} \nabla \times \mathbf{E} &= i\omega \mathbf{B} \\ \nabla \times \mathbf{H} &= -i\omega \mathbf{D} + \mathbf{J} \\ \nabla \cdot \mathbf{B} &= 0 \\ \nabla \cdot \mathbf{D} &= \rho \end{aligned} \qquad (1.11)$$

and in a source-free region, Maxwell equations become

$$\begin{aligned} \nabla \times \mathbf{E} &= i\omega \mathbf{B} \\ \nabla \times \mathbf{H} &= -i\omega \mathbf{D} \\ \nabla \cdot \mathbf{B} &= 0 \\ \nabla \cdot \mathbf{D} &= 0 \end{aligned} \qquad (1.12)$$

The electromagnetic field vectors are linearly related by Maxwell equations, so they all must have the term $e^{i\mathbf{k}\cdot\mathbf{r}}$ (as we see in $\mathbf{E}$ to satisfy the plane wave field characteristics).

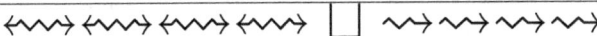

$$
\begin{aligned}
\nabla \times \mathbf{E} &= \nabla \times (\mathbf{E}_0 e^{i\mathbf{k}\cdot\mathbf{r}}) \\
&= e^{i\mathbf{k}\cdot\mathbf{r}}(\nabla \times \mathbf{E}_0) + (\nabla e^{i\mathbf{k}\cdot\mathbf{r}}) \times \mathbf{E}_0 \\
&= i\mathbf{k} \times \mathbf{E} \qquad (1.13)
\end{aligned}
$$

$\mathbf{E}_0$ is independent from $\mathbf{r}$ $\Longrightarrow$ $\nabla \times \mathbf{E}_0 = 0$
where, $\nabla = \frac{d}{dr}\hat{\mathbf{r}}$.

Substituting (1.12) in (1.11) gives

$$i\mathbf{k} \times \mathbf{E} = i\omega\mathbf{B} \Longrightarrow \mathbf{k} \times \mathbf{E}_0 = \omega\mathbf{B}_0 \qquad (1.14)$$

And by doing the same with **D**, **H** and **B**

$$-\mathbf{k} \times \mathbf{H}_0 = \omega\mathbf{D}_0 \qquad (1.15)$$

$$\mathbf{k} \cdot \mathbf{D}_0 = \mathbf{k} \cdot \mathbf{B}_0 = 0 \qquad (1.16)$$

1.4. Laws of Reflection and Refraction

The effects of a boundary surface on the propagating wave is due to the abrupt change in physical properties which literally split the wave into a transmitting part (proceeding into the second medium) and a reflecting part (propagating back into the first medium). These waves must, hence, satisfy the linear boundary conditions at the interface assuming there are no surface charge and current densities there:

$$
\begin{aligned}
\mathbf{E}_I \times \hat{\mathbf{q}} &= \mathbf{E}_{II} \times \hat{\mathbf{q}} \qquad (1.17) \\
\mathbf{H}_I \times \hat{\mathbf{q}} &= \mathbf{H}_{II} \times \hat{\mathbf{q}} \qquad (1.18) \\
\mathbf{B}_I \cdot \hat{\mathbf{q}} &= \mathbf{B}_{II} \cdot \hat{\mathbf{q}} \qquad (1.19) \\
\mathbf{D}_I \cdot \hat{\mathbf{q}} &= \mathbf{D}_{II} \cdot \hat{\mathbf{q}} \qquad (1.20)
\end{aligned}
$$

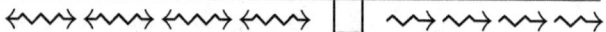

where $\hat{\mathbf{q}}$ is a unit normal to the interface pointing from the first to the second medium. The subscripts I and II denote the total complex field vectors in the first and second medium respectively.

If we take a fixed origin 0 and locate it on the interface and draw the position vector $\mathbf{r}$ from that origin to any point in space, then the vector

$$\widehat{\mathbf{q}} \times \mathbf{r} = \mathbf{r}_p \tag{1.21}$$

defines the position vector of a point on the interface. In other words,the boundary conditions hold only for points $\mathbf{r}_p$. We also note that $\widehat{\mathbf{q}} \times \mathbf{r} = \sin\theta|r|$ or $= \cos\beta|r|$, where theta is the angle between $\widehat{\mathbf{q}}$ and $\mathbf{r}$ and β is the angle equal to $90 - \theta$. The latter gives the projection of the vector on the interface as shown earlier in Figure 1.2.

The electric field strength of the incident wave which exists in medium I is denoted by

$$\mathbf{E}_i = \mathbf{E}_{0i} \exp(i\mathbf{k}_i \cdot \mathbf{r}) \tag{1.22}$$

and the reflected electric field by

$$\mathbf{E}_{r} = \mathbf{E}_{0r} \exp(i\mathbf{k}_{r} \cdot \mathbf{r}) \tag{1.23}$$

and the transmitted electric field by

$$\mathbf{E}_t = \mathbf{E}_{0t} \exp(i\mathbf{k}_t \cdot \mathbf{r}) \tag{1.24}$$

When substituting the electric field strength of the incident, reflected and transmitted waves into the boundary condition (1.16), we get:

$$\begin{aligned}(\mathbf{E}_{0i} \times \widehat{\mathbf{q}}) \exp(i\mathbf{k}_i \cdot \mathbf{r}) &+ (\mathbf{E}_{0r} \times \widehat{\mathbf{q}}) \exp(i\mathbf{k}_{r} \cdot \mathbf{r}) \\ &- (\mathbf{E}_{0t} \times \widehat{\mathbf{q}}) \exp(i\mathbf{k}_t \cdot \mathbf{r}) \\ &= 0 \end{aligned} \tag{1.25}$$

which must be true for all the points $\mathbf{r}_p$ on the interface. And since the vector coefficients of the exponential factors are constant vectors, equation (1.25) contains a linear relation between the exponential functions.

Equation (1.25) shows that either all the vector coefficients are to vanish and be zero $\mathbf{E}_{0i} = \mathbf{E}_{0r} = \mathbf{E}_{0t} = 0$ or that all the exponents $\exp(i\mathbf{k}_i \cdot \mathbf{r})$,$\exp(i\mathbf{k}_{r} \cdot \mathbf{r})$,$\exp(i\mathbf{k}_t \cdot \mathbf{r})$ are equal at all points on the interface. The first case can not exist because $\mathbf{E}_{0i}$ is there already and is not equal to zero. Therefore, we have:

$$\mathbf{k}_i \cdot \mathbf{r}_p = \mathbf{k}_{r} \cdot \mathbf{r}_p = \mathbf{k}_t \cdot \mathbf{r}_p \tag{1.26}$$

for all $\mathbf{r}_p$ on the interface. And if we were to consider a vertical position vector originating from the interface to describe a possible separation between one interface and another (as will be discussed in section 2.3 and 3.8), we introduce

$$\mathbf{r}_v = \widehat{\mathbf{q}} \cdot \mathbf{r} \tag{1.27}$$

with

$$\mathbf{r} = \mathbf{r}_v + \mathbf{r}_p \tag{1.28}$$

where $\mathbf{r}_v$ and $\mathbf{r}_p$ are perpendicular to each others.

By replacing $\mathbf{r}_p$ in (1.25) with $\widehat{\mathbf{q}} \cdot \mathbf{r}$, and by interchanging the positions of the dot and cross in the scalar triple product, and by noting that $\mathbf{r}$ is arbitrary, we obtain:

$$\begin{aligned}
(\mathbf{k}_i - \mathbf{k}_r) \times \widehat{\mathbf{q}} &= \underline{0} = (\mathbf{k}_i - \mathbf{k}_t) \times \widehat{\mathbf{q}} \\
\mathbf{k}_i \times \widehat{\mathbf{q}} &= \mathbf{k}_r \times \widehat{\mathbf{q}} = \mathbf{k}_t \times \widehat{\mathbf{q}} && (1.29) \\
&= \mathbf{a} && (1.30) \\
|a| &= k_i sin\theta_i = k_r sin\theta_r = k_t sin\theta_t && (1.31)
\end{aligned}$$

The last equation expresses the general laws of reflection and refraction as it states that across the interface of any two linear, homogeneous media, the tangential components of the wave vectors of the incident, reflected and transmitted waves, must all be equal.

Having $\mathbf{r}$ as an arbitrary position vector helps us to define the plane of incidence as

$$\mathbf{a} \cdot \mathbf{r} = 0 \tag{1.32}$$

where $\widehat{\mathbf{q}}$ and wave vectors of incident, reflected and transmitted waves all lie on.

The plane of interface is defined using the following relation

$$\widehat{\mathbf{q}} \cdot \mathbf{r} = 0 \tag{1.33}$$

and as shown in Figure 1.3 below.

The geometrical construction of the wave vectors is achieved using

$$\begin{aligned}
\widehat{\mathbf{q}} \times \mathbf{a} &= \mathbf{k}_i - (\mathbf{k}_i \cdot \widehat{\mathbf{q}})\widehat{\mathbf{q}} \\
&= \mathbf{k}_r - (\mathbf{k}_r \cdot \widehat{\mathbf{q}})\widehat{\mathbf{q}} \\
&= \mathbf{k}_t - (\mathbf{k}_t \cdot \widehat{\mathbf{q}})\widehat{\mathbf{q}}
\end{aligned}$$

or

$$\begin{aligned}
\mathbf{k}_i &= \mathbf{b} + q_i\widehat{\mathbf{q}} \\
\mathbf{k}_r &= \mathbf{b} + q_r\widehat{\mathbf{q}} \\
\mathbf{k}_t &= \mathbf{b} + q_t\widehat{\mathbf{q}}
\end{aligned} \tag{1.34}$$

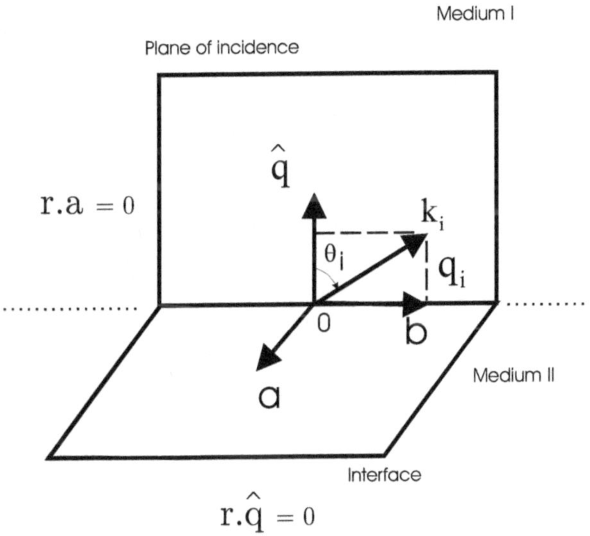

Figure 1.3.: The Orientations of the Interface and the Plane of Incidence.

where

$$\mathbf{b} = \hat{\mathbf{q}} \times \mathbf{a} \tag{1.35}$$

And the projections of the wave vectors on the interface are equal to the constant

$$|\mathbf{a}| = |\mathbf{b}| \tag{1.36}$$

Part II.

ISOTROPIC CASE

2

Plane Waves in Isotropic Media

2.1. Uniform Plane Waves in Linear, Isotropic, Homogeneous, Nonmagnetic and Lossless Media

The relations which describe the behavior of the medium under influence[1] of the electromagnetic field are called *constitutive relations* and can be established using experimentation.

These relations in lossless isotropic medium are :

$$\begin{aligned} \mathbf{D}_0 &= \varepsilon_0 \varepsilon \mathbf{E}_0 \\ \mathbf{B}_0 &= \mu_0 \mu \mathbf{H}_0 \end{aligned} \tag{2.1}$$

where μ and ε are real constants for linear, isotropic, homogeneous, lossless (i.e. instantaneously reacting) and locally reacting material. Substituting these equations into (1.13) and (1.14) gives:

$$\mathbf{k} \times \frac{\mathbf{D}_0}{\varepsilon_0 \varepsilon} = \omega \mathbf{B}_0 \tag{2.2}$$

$$-\mathbf{k} \times \frac{\mathbf{B}_0}{\mu_0 \mu} = \omega \mathbf{D}_0 \tag{2.3}$$

[1] See Appendix F.

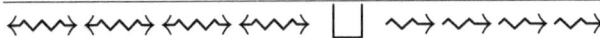

Substituting equation (2.2) into (2.3) gives:

$$-\mathbf{k} \times \frac{\mathbf{k} \times \frac{\mathbf{D}_0}{\varepsilon_0 \varepsilon}}{\omega \mu_0 \mu} = \omega \mathbf{D}_0 \tag{2.4}$$

while taking into consideration that

$$\mathbf{A} \times (\mathbf{B} \times \mathbf{C}) = \mathbf{B}(\mathbf{A} \cdot \mathbf{C}) - \mathbf{C}(\mathbf{A} \cdot \mathbf{B})$$

This turns equation (2.4) into:

$$\frac{\mathbf{D}_0 \mathbf{k}^2}{\omega^2 \varepsilon_0 \varepsilon \mu_0 \mu} = \mathbf{D}_0 \tag{2.5}$$

We also have

$$(n^2 - \varepsilon\mu)\mathbf{E}_0 = 0 \tag{2.6}$$

and similarly

$$(n^2 - \varepsilon\mu)\mathbf{H}_0 = 0 \tag{2.7}$$

The homogeneous equations (2.6) and (2.7) have a nonzero solution ($\mathbf{E}_0$ or $\mathbf{H}_0$) only if the coefficient vanishes, i.e. if

$$n^2 = \varepsilon\mu \tag{2.8}$$

$$\text{or} \quad k^2 = {k_0}^2 \varepsilon\mu = \frac{\omega^2 \varepsilon\mu}{c^2} \tag{2.9}$$

Equation (2.9) is called *Dispersion Equation*, where $k_0 = \omega\sqrt{\varepsilon_0 \mu_0} = \frac{\omega}{c}$.

In this case we have,

$$\mathbf{H}_0 = \frac{1}{\eta}\,(\widehat{\mathbf{k}} \times \mathbf{E}_0) \tag{2.10}$$

$$\mathbf{E}_0 = -\eta\,(\widehat{\mathbf{k}} \times \mathbf{H}_0) \tag{2.11}$$

where η is the intrinsic impedance defined by

$$\eta = \sqrt{\frac{\mu_0 \mu}{\varepsilon_0 \varepsilon}} \tag{2.12}$$

The monochromatic plane wave in a lossless, isotropic medium is a transverse electromagnetic wave, i.e. the electric and magnetic field vectors are perpendicular to each other in this case and to the direction of wave propagation giving the relations in time domain:

$$\hat{\mathbf{k}} \cdot \underline{H} = \hat{\mathbf{k}} \cdot \underline{E} \tag{2.13}$$

$$= \underline{E} \cdot \underline{H} = zero \tag{2.14}$$

2.2. Reflection and Transmission of Waves at Plane Interface

Taking two homogeneous isotropic media separated by an infinite plane interface (as illustrated in Figure 2.1) has a certain effect of discontinuity upon the propagating wave at that surface.

The wave vectors of the reflected waves in both media (1 and 2) must satisfy the above dispersion equation (2.9)

$$\mathbf{k}_r^2 = k_0^2 \mu_1 \varepsilon_1 \tag{2.15}$$

$$\mathbf{k}_t^2 = k_0^2 \mu_2 \varepsilon_2 \tag{2.16}$$

From equation set (1.33) we write

$$q_r = \mathbf{k}_r \cdot \hat{\mathbf{q}} \tag{2.17}$$

And by taking the dot product of equation (1.33) with itself we get

$$q_r^2 = \mathbf{k}_r^2 - \mathbf{a}^2 \tag{2.18}$$

$$\begin{aligned} q_i^2 &= \mathbf{k}_i^2 - \mathbf{a}^2 \\ &= \mathbf{k}_i^2 - \mathbf{k}_i^2 \sin \theta_i^2 \\ &= \mathbf{k}_i^2 (1 - \sin \theta_i^2) = \mathbf{k}_i^2 \cos \theta_i^2 \end{aligned} \tag{2.19}$$

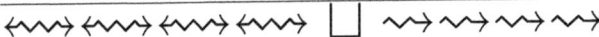

Since both the reflected and the incident waves are in medium I, they satisfy the same dispersion equation (2.15). Thus, equations (2.18) and (2.19) have the following relation:

$$q_i = \pm q_r \tag{2.20}$$

which means two waves can exist in isotropic medium I. The solution $q_i = +q_r$ gives $\mathbf{k}_r = \mathbf{k}_i$ which is the incident wave. The other solution $q_i = -q_r$, and from (2.17)

$$(\mathbf{k}_i + \mathbf{k}_r) \cdot \widehat{\mathbf{q}} = 0 \tag{2.21}$$

leads to

$$(\widehat{\mathbf{k}}_i \cdot \widehat{\mathbf{q}}) = -(\widehat{\mathbf{k}}_r \cdot \widehat{\mathbf{q}}) \tag{2.22}$$

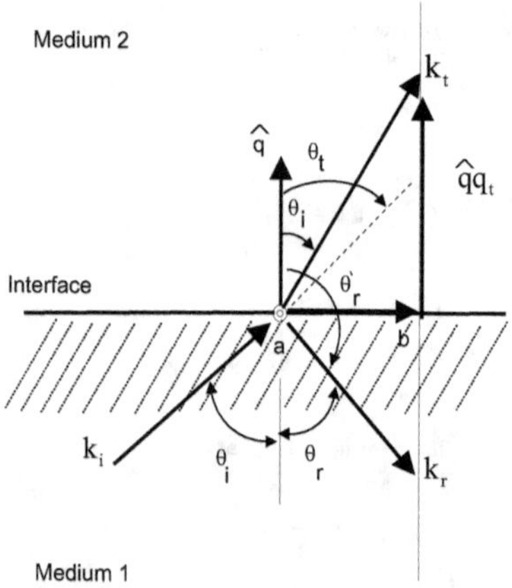

Figure 2.1.: Orientation of the Wave Vectors with Respect to the Normal of the Interface.

And as we see in Figure 2.1 below, it is equivalent to:

$$cos\theta_r' = -cos\theta_i \tag{2.23}$$

And from (1.30)

$$sin\theta_r' = sin\theta_i \tag{2.24}$$

$$\text{Hence,} \quad \theta_i = \pi - \theta_r' \tag{2.25}$$

Similarly,

$$q_t^2 = \mathbf{k}_t^2 - \mathbf{a}^2 = k_0^2\mu_2\varepsilon_2 - \mathbf{a}^2 \tag{2.26}$$

and hence

$$q_t = \pm\sqrt{\mathbf{k}_t^2 - \mathbf{a}^2} \tag{2.27}$$

From (1.28) to (1.30) we get

$$q_t = \pm\sqrt{k_t^2 - k_i^2 sin^2\theta_i} \tag{2.28}$$

(-) sign means that k_t is propagating towards the interface to the first medium, i.e. it's not a transmitted wave. Therefore, q_t must be positive. And from Snell's Law of refraction equation (1.10), we say that the second medium is optically denser than the first if $n_t > n_i$ or $n_{21} > 1$. In this case

$$sin\theta_t = \frac{1}{n_{21}} sin\theta_i < sin\theta_i \tag{2.29}$$

so that $\theta_t < \theta_i$.

If the second medium is optically less dense than the first (i.e. $n_{21} < 1$) we obtain real value for θ_t only for those angles of incidence where $sin\theta_i < n_{21}$. However, for larger values of θ_i, total reflection takes place.

The wave vectors are

$$\mathbf{k}_r = \mathbf{b} - (\mathbf{k}_i \cdot \widehat{\mathbf{q}})\widehat{\mathbf{q}} \tag{2.30}$$

$$\mathbf{k}_t = \mathbf{b} + \sqrt{k_0^2\mu_2\varepsilon_2 - \mathbf{a}^2\widehat{\mathbf{q}}} \tag{2.31}$$

The wave vectors may also be constructed geometrically. According to equation (1.29), we may write

$$\mathbf{k}_r = \mathbf{k}_i + \alpha_r\widehat{\mathbf{q}} \tag{2.32}$$

$$\mathbf{k}_t = \mathbf{k}_i + \alpha_t\widehat{\mathbf{q}} \tag{2.33}$$

where α_r and α_t are two arbitrary constants. The last two equations say that the tips of the wave vectors $\mathbf{k}_t$ and $\mathbf{k}_r$, drawn from fixed origin 0 on the interface, must lie on a straight line which passes through the tip of the given wave vector $\mathbf{k}_i$ and is parallel to $\widehat{\mathbf{q}}$ as seen in Figure 2.2 below.

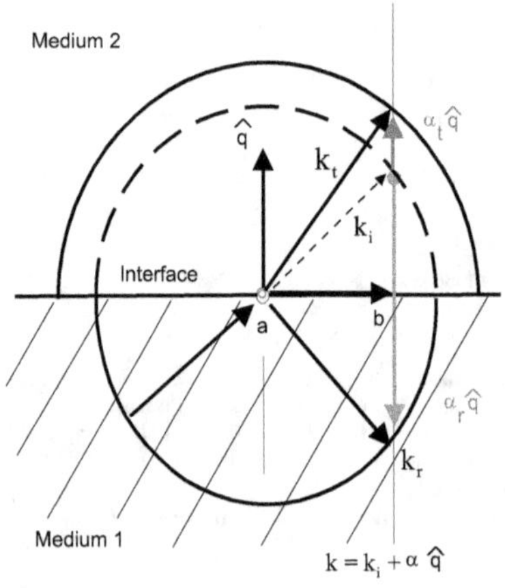

Figure 2.2.: Geometrical Determination of Wave Vectors in Two Isotropic Media. The Vector $\underline{\mathbf{a}}$ Points out of the Paper.

The incident uniform wave is

$$\begin{aligned}\mathbf{E}_i &= \mathbf{E}_{0i} e^{i\mathbf{k}_i \cdot \mathbf{r}} \\ \mathbf{H}_i &= \mathbf{H}_{0i} e^{i\mathbf{k}_i \cdot \mathbf{r}} = \frac{1}{\eta_i} \left(\widehat{\mathbf{k}}_i \times \mathbf{E}_i \right)\end{aligned} \tag{2.34}$$

And the reflected wave is

$$\begin{aligned} \mathbf{E}_r &= \mathbf{E}_{0r} e^{i\mathbf{k}_r \cdot \mathbf{r}} \\ \mathbf{H}_r &= \mathbf{H}_{0r} e^{i\mathbf{k}_r \cdot \mathbf{r}} = \frac{1}{\eta_i} \left(\widehat{\mathbf{k}}_r \times \mathbf{E}_r\right) \end{aligned} \tag{2.35}$$

And the transmitted wave is

$$\begin{aligned} \mathbf{E}_t &= \mathbf{E}_{0t} e^{i\mathbf{k}_t \cdot \mathbf{r}} \\ \mathbf{H}_t &= \mathbf{H}_{0t} e^{i\mathbf{k}_t \cdot \mathbf{r}} = \frac{1}{\eta_t} \left(\widehat{\mathbf{k}}_t \times \mathbf{E}_t\right) \end{aligned} \tag{2.36}$$

where η_i and η_t are the intrinsic impedances of medium 1 and 2 respectively.

According to equation (1.17)

$$(\mathbf{E}_{0i} + \mathbf{E}_{0r} - \mathbf{E}_{0t}) \times \widehat{\mathbf{q}} = \mathbf{0} \tag{2.37}$$

$$(\mathbf{H}_{0i} + \mathbf{H}_{0r} - \mathbf{H}_{0t}) \times \widehat{\mathbf{q}} = \mathbf{0} \tag{2.38}$$

When media 1 and media 2 have the same magnetic properties i.e., if $\mu_1 = \mu_2 = \mu$, the boundary condition (1.19) becomes

$$(\mathbf{H}_{0i} + \mathbf{H}_{0r} - \mathbf{H}_{0t}) \cdot \widehat{\mathbf{q}} = 0 \tag{2.39}$$

Equations (2.38) and (2.39) show that $(\mathbf{H}_{0i}+\mathbf{H}_{0r}-\mathbf{H}_{0t})$ is simultaneously parallel and perpendicular to the unit vector $\widehat{\mathbf{q}}$. This is possible only if the vector is a zero vector, i.e.,

$$\mathbf{H}_{0i} + \mathbf{H}_{0r} - \mathbf{H}_{0t} = \mathbf{0} \tag{2.40}$$

$\mathbf{E}_0$ in a uniform plane wave is perpendicular to its direction of propagation $\widehat{\mathbf{k}}$. We may thus decompose it into a component normal to the plane of incidence called *perpendicular polarization* and a second component lying in the plane of incidence called *parallel polarization*. Therefore

$$\mathbf{E}_{0i} = A_\perp \mathbf{a} + A_\parallel (\widehat{\mathbf{k}}_i \times \mathbf{a}) \tag{2.41}$$

$$\mathbf{E}_{0r} = B_\perp \mathbf{a} + B_\parallel (\widehat{\mathbf{k}}_r \times \mathbf{a}) \tag{2.42}$$

$$\mathbf{E}_{0t} = C_\perp \mathbf{a} + C_\parallel (\widehat{\mathbf{k}}_t \times \mathbf{a}) \tag{2.43}$$

After matching the boundary conditions at the interface, we obtain a linear relationship between the components $B_\parallel$, $B_\perp$ and $A_\parallel$, $A_\perp$. And in matrix form,

$$\begin{bmatrix} B_\perp \\ B_\parallel \end{bmatrix} = \overline{\Gamma} \cdot \begin{bmatrix} A_\perp \\ A_\parallel \end{bmatrix} \tag{2.44}$$

where

$$\overline{\Gamma} = \begin{bmatrix} \Gamma_{11} & \Gamma_{12} \\ \Gamma_{21} & \Gamma_{22} \end{bmatrix} \tag{2.45}$$

is the reflection coefficient matrix, and the entries are defined by

$$\Gamma_{11} = \left.\frac{B_\perp}{A_\perp}\right|_{A_\parallel=0} \qquad \Gamma_{12} = \left.\frac{B_\perp}{A_\parallel}\right|_{A_\perp=0} \tag{2.46}$$

$$\Gamma_{21} = \left.\frac{B_\parallel}{A_\perp}\right|_{A_\parallel=0} \qquad \Gamma_{22} = \left.\frac{B_\parallel}{A_\parallel}\right|_{A_\perp=0} \tag{2.47}$$

And similarly the other components

$$\begin{bmatrix} C_\perp \\ C_\parallel \end{bmatrix} = \overline{T} \cdot \begin{bmatrix} A_\perp \\ A_\parallel \end{bmatrix} \tag{2.48}$$

where

$$\overline{T} = \begin{bmatrix} T_{11} & T_{12} \\ T_{21} & T_{22} \end{bmatrix} \tag{2.49}$$

is the transmission coefficient matrix, and the entries are defined by

$$T_{11} = \left.\frac{C_\perp}{A_\perp}\right|_{A_\parallel=0} \qquad T_{12} = \left.\frac{C_\perp}{A_\parallel}\right|_{A_\perp=0} \tag{2.50}$$

$$T_{21} = \left.\frac{C_\parallel}{A_\perp}\right|_{A_\parallel=0} \qquad T_{22} = \left.\frac{C_\parallel}{A_\parallel}\right|_{A_\perp=0} \tag{2.51}$$

2.2.1. Fresnel Equations when $\mu_1 = \mu_2$

To calculate the reflection and transmission coefficients for two isotropic media of the same permeability, we check the boundary conditions first. From (2.40) we get

$$\mathbf{k}_i \times \mathbf{E}_{0i} + \mathbf{k}_r \times \mathbf{E}_{0r} - \mathbf{k}_t \times \mathbf{E}_{0t} = \mathbf{0} \tag{2.52}$$

Then by taking the dot product of (2.52) with $\mathbf{k}_r$, $\mathbf{k}_t$ and $\mathbf{a}$ respectively, we get

$$(\mathbf{k}_r \times \mathbf{k}_i) \cdot \mathbf{E}_{0i} - (\mathbf{k}_r \times \mathbf{k}_t) \cdot \mathbf{E}_{0t} = 0 \qquad (2.53)$$

$$(\mathbf{k}_t \times \mathbf{k}_i) \cdot \mathbf{E}_{0i} + (\mathbf{k}_t \times \mathbf{k}_r) \cdot \mathbf{E}_{0r} = 0 \qquad (2.54)$$

$$(\mathbf{a} \times \mathbf{k}_i) \cdot \mathbf{E}_{0i} + (\mathbf{a} \times \mathbf{k}_r) \cdot \mathbf{E}_{0r} - (\mathbf{a} \times \mathbf{k}_t) \cdot \mathbf{E}_{0t} = 0 \qquad (2.55)$$

where

$$(\mathbf{A} \times \mathbf{B}) \cdot \mathbf{C} = \mathbf{A} \cdot (\mathbf{B} \times \mathbf{C})$$

Equations (2.41), (2.42) and (2.43) are to be substituted into equations (2.53), (2.54) and (2.55). And also noting that $(\mathbf{a} \times \widehat{\mathbf{k}}_\alpha)^2 = \mathbf{a}^2$ where α denotes i, r or t, we obtain

$$(\mathbf{a} \cdot \mathbf{k}_r \times \mathbf{k}_i)A_\perp - (\mathbf{a} \cdot \mathbf{k}_r \times \mathbf{k}_t)C_\perp = 0 \qquad (2.56)$$

$$(\mathbf{a} \cdot \mathbf{k}_t \times \mathbf{k}_i)A_\perp + (\mathbf{a} \cdot \mathbf{k}_t \times \mathbf{k}_r)B_\perp = 0 \qquad (2.57)$$

$$k_i A_\parallel + k_r B_\parallel - k_t C_\parallel = 0 \qquad (2.58)$$

where

$$(\mathbf{a} \times \mathbf{b}) \cdot (\mathbf{c} \times \mathbf{d}) = (\mathbf{a} \cdot \mathbf{c})(\mathbf{b} \cdot \mathbf{d}) - (\mathbf{a} \cdot \mathbf{d})(\mathbf{b} \cdot \mathbf{c})$$

From (2.56) and (2.57) we get

$$\frac{A_\perp}{\mathbf{a} \cdot (\mathbf{k}_r \times \mathbf{k}_t)} = \frac{B_\perp}{\mathbf{a} \cdot (\mathbf{k}_t \times \mathbf{k}_i)} = \frac{C_\perp}{\mathbf{a} \cdot (\mathbf{k}_r \times \mathbf{k}_i)} \qquad (2.59)$$

We can also derive general relations to connect the three wave vectors:

$$\mathbf{k}_\alpha \times \mathbf{k}_\beta = (\mathbf{b} + q_\alpha \widehat{\mathbf{q}}) \times (\mathbf{b} + q_\beta \widehat{\mathbf{q}}) \qquad (2.60)$$

$$= (\mathbf{b} \times q_\beta \widehat{\mathbf{q}}) + (q_\alpha \widehat{\mathbf{q}} \times \mathbf{b}) \qquad (2.61)$$

$$= (q_\beta - q_\alpha)\mathbf{a} \qquad (2.62)$$

where $\mathbf{b} \times \widehat{\mathbf{q}} = \mathbf{a}$. And if (2.62) to be substituted in:

$$\mathbf{k}_\alpha \times \mathbf{k}_\beta + \mathbf{k}_\beta \times \mathbf{k}_\gamma + \mathbf{k}_\gamma \times \mathbf{k}_\alpha = 0 \qquad (2.63)$$

$$q_\alpha(\mathbf{k}_\beta \times \mathbf{k}_\gamma) + q_\beta(\mathbf{k}_\gamma \times \mathbf{k}_\alpha) + q_\gamma(\mathbf{k}_\alpha \times \mathbf{k}_\beta) = 0 \qquad (2.64)$$

$$(\mathbf{k}_\beta \times \mathbf{k}_\gamma)\mathbf{k}_\alpha + (\mathbf{k}_\gamma \times \mathbf{k}_\alpha)\mathbf{k}_\beta + (\mathbf{k}_\alpha \times \mathbf{k}_\beta)\mathbf{k}_\gamma = \bar{0} \qquad (2.65)$$

then the last equations has been proved to be true.

Equation (2.65) is a dyadic equation and follows from the fact that, $q_\alpha = (\mathbf{k}_\alpha - \mathbf{b}) \cdot \widehat{\mathbf{q}}$
where

$$\mathbf{0} = \begin{bmatrix} 0 & 0 & 0 \end{bmatrix} \quad (2.66)$$

$$\overline{\mathbf{0}} = \begin{bmatrix} 0 & 0 & 0 \\ 0 & 0 & 0 \\ 0 & 0 & 0 \end{bmatrix} \quad (2.67)$$

From(2.59), it implies that

$$A_\perp = B_\perp(\frac{q_t - q_r}{q_i - q_t})$$

$$C_\perp = B_\perp(\frac{q_i - q_r}{q_i - q_t})$$

giving

$$\begin{aligned} A_\perp + B_\perp &- C_\perp \\ &= B_\perp(\frac{q_t - q_r}{q_i - q_t}) + B_\perp \\ &\quad -B_\perp(\frac{q_i - q_r}{q_i - q_t}) \\ &= 0 \end{aligned} \quad (2.68)$$

And by taking into consideration equation (1.34)

$$\begin{aligned} A_\perp \mathbf{k}_i + B_\perp \mathbf{k}_r &- C_\perp \mathbf{k}_t \\ &= \mathbf{k}_i B_\perp(\frac{q_t - q_r}{q_i - q_t}) + \mathbf{k}_r B_\perp \\ &\quad -\mathbf{k}_t B_\perp(\frac{q_i - q_r}{q_i - q_t}) \\ &= (\mathbf{b} + q_i\widehat{\mathbf{q}})B_\perp(\frac{q_t - q_r}{q_i - q_t}) + (\mathbf{b} + q_r\widehat{\mathbf{q}})B_\perp \\ &\quad -(\mathbf{b} + q_t\widehat{\mathbf{q}})B_\perp(\frac{q_i - q_r}{q_i - q_t}) \\ &= \mathbf{0} \end{aligned} \quad (2.69)$$

Dot multiplying (2.69) by $\widehat{\mathbf{k}}_t$ and $(\mathbf{a} \times \mathbf{k}_t)$ respectively yields

$$k_i(\widehat{\mathbf{k}}_i \cdot \widehat{\mathbf{k}}_t)A_\perp + k_r(\widehat{\mathbf{k}}_r \cdot \widehat{\mathbf{k}}_t)B_\perp - k_t C_\perp = 0 \quad (2.70)$$

$$(\widehat{\mathbf{k}}_i \cdot \widehat{\mathbf{q}})(\widehat{\mathbf{k}}_i \cdot \widehat{\mathbf{k}}_t)A_\perp + (\widehat{\mathbf{k}}_r \cdot \widehat{\mathbf{q}})(\widehat{\mathbf{k}}_r \cdot \widehat{\mathbf{k}}_t)B_\perp - (\widehat{\mathbf{k}}_t \cdot \widehat{\mathbf{q}})C_\perp = 0 \quad (2.71)$$

The angle of incidence lies within the range of 0 to 180 degrees where the sine is always positive, which means I do not need absolute values in

$$\mathbf{a} = \mathbf{k}_i \times \widehat{\mathbf{q}} = \mathbf{k}_r \times \widehat{\mathbf{q}} = \mathbf{k}_t \times \widehat{\mathbf{q}}$$

And equation

$$\begin{aligned}
(\mathbf{a} \times \mathbf{k}_t) \cdot \mathbf{k}_\alpha &= (\mathbf{k}_\alpha \times \widehat{\mathbf{q}} \times \mathbf{k}_t) \cdot \mathbf{k}_\alpha \\
&= ((\widehat{\mathbf{q}}\mathbf{k}_\alpha - \mathbf{k}_\alpha \widehat{\mathbf{q}}) \cdot \mathbf{k}_t) \cdot \mathbf{k}_\alpha \\
&= (\widehat{\mathbf{q}}\mathbf{k}_\alpha \cdot \mathbf{k}_t - \mathbf{k}_\alpha \widehat{\mathbf{q}} \cdot \mathbf{k}_t) \cdot \mathbf{k}_\alpha \\
&= (\widehat{\mathbf{q}} \cdot \mathbf{k}_\alpha)(\mathbf{k}_\alpha \cdot \mathbf{k}_t) - (\mathbf{k}_\alpha \cdot \mathbf{k}_\alpha)(\widehat{\mathbf{q}} \cdot \mathbf{k}_t)
\end{aligned}$$

has been used in calculating (2.71), so that

$$\begin{aligned}
A_\perp(\widehat{\mathbf{q}} \cdot \mathbf{k}_i)(\mathbf{k}_i \cdot \mathbf{k}_t) &- A_\perp(\mathbf{k}_i \cdot \mathbf{k}_i)(\widehat{\mathbf{q}} \cdot \mathbf{k}_t) \\
+B_\perp(\widehat{\mathbf{q}} \cdot \mathbf{k}_r)(\mathbf{k}_r \cdot \mathbf{k}_t) &- B_\perp(\mathbf{k}_r \cdot \mathbf{k}_r)(\widehat{\mathbf{q}} \cdot \mathbf{k}_t) \\
-C_\perp(\widehat{\mathbf{q}} \cdot \mathbf{k}_t)(\mathbf{k}_t \cdot \mathbf{k}_t) &+ C_\perp(\mathbf{k}_t \cdot \mathbf{k}_t)(\widehat{\mathbf{q}} \cdot \mathbf{k}_t) \\
&= 0 \\
= A_\perp(\widehat{\mathbf{q}} \cdot \mathbf{k}_i)(\mathbf{k}_i \cdot \mathbf{k}_t) &- A_\perp(\mathbf{k}_i \cdot \mathbf{k}_i)(\widehat{\mathbf{q}} \cdot \mathbf{k}_t) \\
+B_\perp(\widehat{\mathbf{q}} \cdot \mathbf{k}_r)(\mathbf{k}_r \cdot \mathbf{k}_t) &- B_\perp(\mathbf{k}_r \cdot \mathbf{k}_r)(\widehat{\mathbf{q}} \cdot \mathbf{k}_t) \\
&\Longrightarrow \\
A_\perp(\widehat{\mathbf{q}} \cdot \widehat{\mathbf{k}}_i)(\widehat{\mathbf{k}}_i \cdot \widehat{\mathbf{k}}_t) &- A_\perp(\widehat{\mathbf{k}}_i \cdot \widehat{\mathbf{k}}_i)(\widehat{\mathbf{q}} \cdot \widehat{\mathbf{k}}_t) \\
+B_\perp(\widehat{\mathbf{q}} \cdot \widehat{\mathbf{k}}_r)(\widehat{\mathbf{k}}_r \cdot \widehat{\mathbf{k}}_t) &- B_\perp(\widehat{\mathbf{k}}_r \cdot \widehat{\mathbf{k}}_r)(\widehat{\mathbf{q}} \cdot \widehat{\mathbf{k}}_t) \\
&= 0
\end{aligned}$$

The latter was divided over $\mathbf{k}_r^2$ or $\mathbf{k}_i^2$ because $|\mathbf{k}_r| = |\mathbf{k}_i|$ and also over $|\mathbf{k}_t|$ in addition to taking into consideration the operation $\widehat{\mathbf{k}}_r \cdot \widehat{\mathbf{k}}_r = 1$ and $\widehat{\mathbf{k}}_i \cdot \widehat{\mathbf{k}}_i = 1$.

With the help of equation (2.68)

$$\begin{aligned}
\Longrightarrow \quad &= A_\perp(\widehat{\mathbf{q}}\cdot\widehat{\mathbf{k}}_i)(\widehat{\mathbf{k}}_i\cdot\widehat{\mathbf{k}}_t) - A_\perp(\widehat{\mathbf{k}}_i\cdot\widehat{\mathbf{k}}_i)(\widehat{\mathbf{q}}\cdot\widehat{\mathbf{k}}_t) \\
&\quad + B_\perp(\widehat{\mathbf{q}}\cdot\widehat{\mathbf{k}}_r)(\widehat{\mathbf{k}}_r\cdot\widehat{\mathbf{k}}_t) - B_\perp(\widehat{\mathbf{k}}_r\cdot\widehat{\mathbf{k}}_r)(\widehat{\mathbf{q}}\cdot\widehat{\mathbf{k}}_t) \\
&= A_\perp(\widehat{\mathbf{q}}\cdot\widehat{\mathbf{k}}_i)(\widehat{\mathbf{k}}_i\cdot\widehat{\mathbf{k}}_t) \\
&\quad + B_\perp(\widehat{\mathbf{q}}\cdot\widehat{\mathbf{k}}_r)(\widehat{\mathbf{k}}_r\cdot\widehat{\mathbf{k}}_t) \\
&\quad - C_\perp(\widehat{\mathbf{q}}\cdot\widehat{\mathbf{k}}_t) \\
&= 0
\end{aligned}$$

Substituting equations (2.41), (2.42) and (2.43) into equation (2.37) and taking into account (2.68) and the fact $(\widehat{\mathbf{k}}_\alpha \times \mathbf{a}) \times \widehat{\mathbf{q}} = (\widehat{\mathbf{k}}_\alpha \cdot \widehat{\mathbf{q}})\mathbf{a}$, we obtain

$$\begin{aligned}
(A_\perp \mathbf{a} + A_\parallel(\widehat{\mathbf{k}}_i \times \mathbf{a}) + B_\perp \mathbf{a} + B_\parallel(\widehat{\mathbf{k}}_r \times \mathbf{a}) & \\
-C_\perp \mathbf{a} - C_\parallel(\widehat{\mathbf{k}}_t \times \mathbf{a})) \times \widehat{\mathbf{q}} &= 0
\end{aligned}$$

$$\begin{aligned}
A_\perp \mathbf{a}\times\widehat{\mathbf{q}} + A_\parallel(\widehat{\mathbf{k}}_i\cdot\widehat{\mathbf{q}})\mathbf{a} + B_\perp \mathbf{a}\times\widehat{\mathbf{q}} + B_\parallel(\widehat{\mathbf{k}}_r\cdot\widehat{\mathbf{q}})\mathbf{a} & \\
-C_\perp \mathbf{a}\times\widehat{\mathbf{q}} - C_\parallel(\widehat{\mathbf{k}}_t\cdot\widehat{\mathbf{q}})\mathbf{a} &= 0
\end{aligned}$$

$$\begin{aligned}
A_\parallel(\widehat{\mathbf{k}}_i\cdot\widehat{\mathbf{q}})\mathbf{a} + B_\parallel(\widehat{\mathbf{k}}_r\cdot\widehat{\mathbf{q}})\mathbf{a} & \\
-C_\parallel(\widehat{\mathbf{k}}_t\cdot\widehat{\mathbf{q}})\mathbf{a} &= 0
\end{aligned}$$

$$\begin{aligned}
(A_\parallel(\widehat{\mathbf{k}}_i\cdot\widehat{\mathbf{q}})\mathbf{a} + B_\parallel(\widehat{\mathbf{k}}_r\cdot\widehat{\mathbf{q}})\mathbf{a} & \\
-C_\parallel(\widehat{\mathbf{k}}_t\cdot\widehat{\mathbf{q}})\mathbf{a} &= 0)\cdot\mathbf{a}
\end{aligned}$$

$$A_\parallel(\widehat{\mathbf{k}}_i\cdot\widehat{\mathbf{q}}) + B_\parallel(\widehat{\mathbf{k}}_r\cdot\widehat{\mathbf{q}}) - C_\parallel(\widehat{\mathbf{k}}_t\cdot\widehat{\mathbf{q}}) = 0 \tag{2.72}$$

Equations (2.68) and (2.72) may be solved for $A_{\|}/C_{\|}$ and $B_{\|}/C_{\|}$. Dividing both equations over each others gives us

$$\begin{aligned}
\frac{k_r B_{\|}}{B_{\|}(\widehat{\mathbf{k}}_r \cdot \widehat{\mathbf{q}})} &= \frac{-k_i A_{\|} + k_t C_{\|}}{-A_{\|}(\widehat{\mathbf{k}}_i \cdot \widehat{\mathbf{q}}) + C_{\|}(\widehat{\mathbf{k}}_t \cdot \widehat{\mathbf{q}})} \\
(k_r)(-A_{\|}(\widehat{\mathbf{k}}_i \cdot \widehat{\mathbf{q}}) + C_{\|}(\widehat{\mathbf{k}}_t \cdot \widehat{\mathbf{q}})) &= (\widehat{\mathbf{k}}_r \cdot \widehat{\mathbf{q}})(-k_i A_{\|} + k_t C_{\|}) \\
(k_r)(\frac{-A_{\|}}{C_{\|}}(\widehat{\mathbf{k}}_i \cdot \widehat{\mathbf{q}}) + (\widehat{\mathbf{k}}_t \cdot \widehat{\mathbf{q}})) &= (\widehat{\mathbf{k}}_r \cdot \widehat{\mathbf{q}})(\frac{-k_i A_{\|}}{C_{\|}} + k_t) \\
\frac{A_{\|}}{C_{\|}} &= \frac{k_r(\widehat{\mathbf{k}}_t \cdot \widehat{\mathbf{q}}) - k_t(\widehat{\mathbf{k}}_r \cdot \widehat{\mathbf{q}})}{k_r(\widehat{\mathbf{k}}_i \cdot \widehat{\mathbf{q}}) - k_i(\widehat{\mathbf{k}}_r \cdot \widehat{\mathbf{q}})}
\end{aligned}$$

or

$$\frac{A_{\|}}{C_{\|}} = \frac{\begin{vmatrix} k_t & k_r \\ \widehat{\mathbf{k}}_t \cdot \widehat{\mathbf{q}} & \widehat{\mathbf{k}}_r \cdot \widehat{\mathbf{q}} \end{vmatrix}}{\Delta} \tag{2.73}$$

and

$$\frac{B_{\|}}{C_{\|}} = \frac{\begin{vmatrix} k_i & k_t \\ \widehat{\mathbf{k}}_i \cdot \widehat{\mathbf{q}} & \widehat{\mathbf{k}}_t \cdot \widehat{\mathbf{q}} \end{vmatrix}}{\Delta} \tag{2.74}$$

where

$$\Delta = \begin{vmatrix} k_i & k_r \\ \widehat{\mathbf{k}}_i \cdot \widehat{\mathbf{q}} & \widehat{\mathbf{k}}_r \cdot \widehat{\mathbf{q}} \end{vmatrix}$$

Similarly, solving equations (2.70) and(2.71) for $A_{\perp}/C_{\perp}$ and $B_{\perp}/C_{\perp}$ and then comparing the results with equations (2.73) and (2.74), we find

$$\frac{A_{\|}/A_{\perp}}{\widehat{\mathbf{k}}_i \cdot \widehat{\mathbf{k}}_t} = \frac{B_{\|}/B_{\perp}}{\widehat{\mathbf{k}}_r \cdot \widehat{\mathbf{k}}_t} = \frac{C_{\|}/C_{\perp}}{\widehat{\mathbf{k}}_t \cdot \widehat{\mathbf{k}}_t} \tag{2.75}$$

where $\widehat{\mathbf{k}}_t \cdot \widehat{\mathbf{k}}_t = 1$. And multiplying (2.75) by $1/k_i k_t$ and noting that $q_r = -q_i$, and

$$\begin{aligned}
\widehat{\mathbf{k}}_\alpha \cdot \widehat{\mathbf{k}}_\beta &= (\mathbf{b} + q_\alpha \widehat{\mathbf{q}}) \cdot (\mathbf{b} + q_\beta \widehat{\mathbf{q}}) \\
&= \mathbf{b}^2 + q_\alpha q_\beta \\
&= \mathbf{a}^2 + q_\alpha q_\beta
\end{aligned}$$

we may rewrite equation (2.75) as

$$\frac{A_{\|}/A_{\perp}}{\mathbf{a}^2 + q_i q_t} = \frac{B_{\|}/B_{\perp}}{\mathbf{a}^2 - q_i q_t} = \frac{C_{\|}/C_{\perp}}{k_i k_t}$$

For a given uniform plane wave incident on an interface of two lossless isotropic media, equations (2.59) and (2.75) determine the reflection and transmission coefficients which are defined in (2.46), (2.47), (2.50) and (2.51), and will become

$$\Gamma_{11} = \left.\frac{B_\perp}{A_\perp}\right|_{A_\parallel=0} = \frac{\mathbf{a}\cdot(\mathbf{k}_t \times \mathbf{k}_i)}{\mathbf{a}\cdot(\mathbf{k}_r \times \mathbf{k}_t)} \tag{2.76}$$

$$\Gamma_{12} = \left.\frac{B_\perp}{A_\parallel}\right|_{A_\perp=0} = 0 \tag{2.77}$$

$$\Gamma_{21} = \left.\frac{B_\parallel}{A_\perp}\right|_{A_\parallel=0} = 0 \tag{2.78}$$

$$\Gamma_{22} = \left.\frac{B_\parallel}{A_\parallel}\right|_{A_\perp=0} = \frac{\widehat{\mathbf{k}}_r \cdot \widehat{\mathbf{k}}_t}{\widehat{\mathbf{k}}_i \cdot \widehat{\mathbf{k}}_t}\,\Gamma_{11} \tag{2.79}$$

and

$$T_{11} = \left.\frac{C_\perp}{A_\perp}\right|_{A_\parallel=0} = \frac{\mathbf{a}\cdot(\mathbf{k}_r \times \mathbf{k}_i)}{\mathbf{a}\cdot(\mathbf{k}_r \times \mathbf{k}_t)} \tag{2.80}$$

$$T_{12} = \left.\frac{C_\perp}{A_\parallel}\right|_{A_\perp=0} = 0 \tag{2.81}$$

$$T_{21} = \left.\frac{C_\parallel}{A_\perp}\right|_{A_\parallel=0} = 0 \tag{2.82}$$

$$T_{22} = \left.\frac{C_\parallel}{A_\parallel}\right|_{A_\perp=0} = \frac{1}{\widehat{\mathbf{k}}_i \cdot \widehat{\mathbf{k}}_t}\,T_{11} \tag{2.83}$$

We may also express the reflection and transmission coefficients in terms of the angle of incidence and the angle of transmission, knowing that the vector $\mathbf{k}_\alpha \times \mathbf{k}_\beta$ is parallel to $\mathbf{a}$. We find:

$$\begin{aligned}
\mathbf{a}\cdot(\mathbf{k}_t \times \mathbf{k}_i) &= |\mathbf{a}||\mathbf{k}_t \times \mathbf{k}_i| = |\mathbf{a}|k_t k_i \sin(\theta_t - \theta_i) \\
\mathbf{a}\cdot(\mathbf{k}_r \times \mathbf{k}_t) &= |\mathbf{a}||\mathbf{k}_r \times \mathbf{k}_t| = |\mathbf{a}|k_t k_i \sin(\theta_t + \theta_i) \\
\mathbf{a}\cdot(\mathbf{k}_r \times \mathbf{k}_i) &= |\mathbf{a}||\mathbf{k}_r \times \mathbf{k}_i| = |\mathbf{a}|k_i^2 \sin 2\theta_i
\end{aligned} \tag{2.84}$$

and

$$\begin{aligned}\widehat{\mathbf{k}}_r \cdot \widehat{\mathbf{k}}_t &= \cos(\theta_r^{'} - \theta_t) \\ &= \cos(180 - \theta_i - \theta_t) \\ &= -\cos(\theta_i + \theta_t) \\ \widehat{\mathbf{k}}_i \cdot \widehat{\mathbf{k}}_t &= = \cos(\theta_i - \theta_t)\end{aligned} \tag{2.85}$$

giving

$$\Gamma_{11} = \frac{\sin(\theta_t - \theta_i)}{\sin(\theta_t + \theta_i)} \tag{2.86}$$

$$\Gamma_{22} = \frac{\tan(\theta_i - \theta_t)}{\tan(\theta_i + \theta_t)} \tag{2.87}$$

$$T_{11} = \frac{2\sin(\theta_t)\cos(\theta_i)}{\sin(\theta_t + \theta_i)} \tag{2.88}$$

$$T_{22} = \frac{2\sin(\theta_t)\cos(\theta_i)}{\sin(\theta_t + \theta_i)\cos(\theta_i - \theta_t)} \tag{2.89}$$

Equations (2.86) to (2.89) are called *Fresnel equations*. Other forms of reflection and transmission coefficients can be derived from

$$\begin{aligned}\mathbf{a} \cdot (\mathbf{k}_t \times \mathbf{k}_i) &= (q_i - q_t)\mathbf{a}^2 \\ \mathbf{a} \cdot (\mathbf{k}_r \times \mathbf{k}_t) &= (q_i + q_t)\mathbf{a}^2 \\ \mathbf{a} \cdot (\mathbf{k}_r \times \mathbf{k}_i) &= 2q_i\mathbf{a}^2\end{aligned} \tag{2.90}$$

and by taking the ratio of equations (2.73) and (2.74)

$$\Gamma_{11} = \frac{q_i - q_t}{q_i + q_t} = \frac{k_i \cos\theta_i - k_t \cos\theta_t}{k_i \cos\theta_i + k_t \cos\theta_t} \tag{2.91}$$

$$\begin{aligned}\Gamma_{22} &= \frac{k_t(\widehat{\mathbf{k}}_i \cdot \widehat{\mathbf{q}}) - k_i(\widehat{\mathbf{k}}_t \cdot \widehat{\mathbf{q}})}{k_t(\widehat{\mathbf{k}}_i \cdot \widehat{\mathbf{q}}) + k_i(\widehat{\mathbf{k}}_t \cdot \widehat{\mathbf{q}})} \\ &= \frac{k_t \cos\theta_i - k_i \cos\theta_t}{k_t \cos\theta_i + k_i \cos\theta_t}\end{aligned} \tag{2.92}$$

$$T_{11} = \frac{2q_i}{q_i + q_t} = \frac{2k_i \cos\theta_i}{k_i \cos\theta_i + k_t \cos\theta_t} \tag{2.93}$$

$$\begin{aligned}T_{22} &= \frac{2k_i(\widehat{\mathbf{k}}_i \cdot \widehat{\mathbf{q}})}{k_i(\widehat{\mathbf{k}}_t \cdot \widehat{\mathbf{q}}) + k_t(\widehat{\mathbf{k}}_t \cdot \widehat{\mathbf{q}})} \\ &= \frac{2k_i \cos\theta_i}{k_i \cos\theta_t + k_t \cos\theta_i}\end{aligned} \tag{2.94}$$

By decomposing the incident wave into perpendicular and parallel components we were able to find the reflection and transmission coefficients. Alternatively, we can also express the amplitude vectors by deriving them directly from the boundary conditions starting with Maxwell's equation:

$$\mathbf{E}_0 = -\frac{1}{\omega\varepsilon_0\varepsilon}(\mathbf{k} \times \mathbf{H}_0) = -\eta(\widehat{\mathbf{k}} \times \mathbf{H}_0) \tag{2.95}$$

And by rewriting equation (2.37)

$$[\mathbf{k}_i \times \mathbf{H}_{0i} + \mathbf{k}_r \times \mathbf{H}_{0r} - \frac{\varepsilon_1}{\varepsilon_2}\mathbf{k}_t \times \mathbf{H}_{0t}] \times \widehat{\mathbf{q}} = \mathbf{0} \tag{2.96}$$

and substituting $\mathbf{H}_{0r} = \mathbf{H}_{0t} - \mathbf{H}_{0i}$ into equation (2.96) and noting that $\mathbf{k}_i - \mathbf{k}_r = 2q_i\widehat{\mathbf{q}}$ we obtain:

$$[(\mathbf{k}_i - \mathbf{k}_r) \times \mathbf{H}_{0i} + (\mathbf{k}_r - \frac{\varepsilon_1}{\varepsilon_2}\mathbf{k}_t) \times \mathbf{H}_{0t}] \times \widehat{\mathbf{q}} = \mathbf{0}$$

$$\begin{aligned} \mathbf{0} = \widehat{\mathbf{q}} &\times [(2q_i\widehat{\mathbf{q}}) \times \mathbf{H}_{0i} + (\mathbf{k}_r - \frac{\varepsilon_1}{\varepsilon_2}\mathbf{k}_t) \times \mathbf{H}_{0t}] \\ \widehat{\mathbf{q}} &\times [(2q_i\widehat{\mathbf{q}}) \times \mathbf{H}_{0i} + (\mathbf{k}_r - \frac{\varepsilon_1}{\varepsilon_2}\mathbf{k}_t) \times \mathbf{H}_{0t}] = \mathbf{0} \\ \widehat{\mathbf{q}} &\times [(2q_i\widehat{\mathbf{q}}) \times \mathbf{H}_{0i} = (\frac{\varepsilon_1}{\varepsilon_2}\mathbf{k}_t - \mathbf{k}_r) \times \mathbf{H}_{0t}] \end{aligned} \tag{2.97}$$

To solve for $\mathbf{H}_{0t}$, we take the cross product of equation (2.97) first with $(\varepsilon_1\mathbf{k}_t/\varepsilon_2 - \mathbf{k}_r)$ and then with $\mathbf{k}_t$. Then we take the right part of the equation:

$$
\begin{aligned}
&= \mathbf{k}_t \times [\mathbf{k}_t \frac{\varepsilon_1}{\varepsilon_2} - \mathbf{k}_r] \times \{\widehat{\mathbf{q}} \times [\frac{\varepsilon_1}{\varepsilon_2}\mathbf{k}_t \times \mathbf{H}_{0t} - \mathbf{k}_r \times \mathbf{H}_{0t}]\} \\
&= \mathbf{k}_t \times [\mathbf{k}_t \frac{\varepsilon_1}{\varepsilon_2} - \mathbf{k}_r] \times \{\widehat{\mathbf{q}} \times (\mathbf{k}_t \times \mathbf{H}_{0t})\frac{\varepsilon_1}{\varepsilon_2} - \widehat{\mathbf{q}} \times (\mathbf{k}_r \times \mathbf{H}_{0t})\} \\
&= \mathbf{k}_t \times [\mathbf{k}_t \frac{\varepsilon_1}{\varepsilon_2} - \mathbf{k}_r] \times \{\mathbf{k}_t(\widehat{\mathbf{q}} \cdot \mathbf{H}_{0t})\frac{\varepsilon_1}{\varepsilon_2} - \mathbf{H}_{0t}(\widehat{\mathbf{q}} \cdot \mathbf{k}_t)\frac{\varepsilon_1}{\varepsilon_2} \\
&\quad - \mathbf{k}_r(\widehat{\mathbf{q}} \cdot \mathbf{H}_{0t}) + \mathbf{H}_{0t}(\widehat{\mathbf{q}} \cdot \mathbf{k}_r)\} \\
&= \mathbf{k}_t \times [\mathbf{k}_t \frac{\varepsilon_1}{\varepsilon_2} - \mathbf{k}_r] \times \{(\widehat{\mathbf{q}} \cdot \mathbf{H}_{0t})[\mathbf{k}_t \frac{\varepsilon_1}{\varepsilon_2} - \mathbf{k}_r] \\
&\quad + \mathbf{H}_{0t}(-\frac{\varepsilon_1}{\varepsilon_2}(\widehat{\mathbf{q}} \cdot \mathbf{k}_t) + (\widehat{\mathbf{q}} \cdot \mathbf{k}_r))\} \\
&= \mathbf{k}_t \times \ \{[(\mathbf{k}_t \times \mathbf{H}_{0t})\frac{\varepsilon_1}{\varepsilon_2} - \mathbf{k}_r \times \mathbf{H}_{0t}](-\frac{\varepsilon_1}{\varepsilon_2} q_t + q_r)\} \\
&= (\mathbf{k}_t \underbrace{(\mathbf{k}_t \cdot \mathbf{H}_{0t})}_{=0} \frac{\varepsilon_1}{\varepsilon_2} - \mathbf{H}_{0t}\mathbf{k}_t^2 \frac{\varepsilon_1}{\varepsilon_2} - \mathbf{k}_r \underbrace{(\mathbf{k}_t \cdot \mathbf{H}_{0t})}_{=0} \\
&\quad + \mathbf{H}_{0t}(\mathbf{k}_t \cdot \mathbf{k}_r))(-\frac{\varepsilon_1}{\varepsilon_2} q_t + q_r) \\
&= (-\mathbf{H}_{0t}\mathbf{k}_t^2 \frac{\varepsilon_1}{\varepsilon_2} + \mathbf{H}_{0t}(\mathbf{k}_t \cdot \mathbf{k}_r))(-\frac{\varepsilon_1}{\varepsilon_2} q_t + q_r)
\end{aligned}
$$

It yields

$$
\begin{aligned}
&2q_i \mathbf{k}_t \times \{(\frac{\varepsilon_1}{\varepsilon_2}\mathbf{k}_t - \mathbf{k}_r) \times [\widehat{\mathbf{q}} \times (\widehat{\mathbf{q}} \times \mathbf{H}_{0i})]\} \\
&\quad = (\frac{\varepsilon_1}{\varepsilon_2} q_t - q_r)(\frac{\varepsilon_1}{\varepsilon_2} k_t^2 - \mathbf{k}_t \cdot \mathbf{k}_r)\mathbf{H}_{0t}
\end{aligned} \tag{2.98}
$$

And by presenting other equations that might help us solve the above equation, we find

$$
\begin{aligned}
(1) \quad \frac{\varepsilon_1}{\varepsilon_2}\mathbf{k}_t^2 - \mathbf{k}_t \cdot \mathbf{k}_r &= \frac{\varepsilon_1}{\varepsilon_2}(k_0^2\varepsilon_2\mu_2) - \mathbf{k}_t \cdot \mathbf{k}_r \\
&= \varepsilon_1(k_0^2\mu_2) - \mathbf{a}^2 - q_r q_t \\
&= k_i^2 - \mathbf{a}^2 - q_r q_t \\
&= \mathbf{a}^2 + q_i^2 - \mathbf{a}^2 - q_r q_t \\
&= q_i(q_i + q_t) \qquad (2.99)
\end{aligned}
$$

$$
\begin{aligned}
(2) \quad \frac{1}{k_t^2}(q_i + q_t)(\mathbf{k}_i \cdot \mathbf{k}_t) &= \frac{1}{k_t^2}(q_i + q_t)(\mathbf{a}^2 + q_i q_t) \\
&= \frac{1}{k_t^2}(q_i + q_t)\mathbf{a}^2 + \frac{1}{k_t^2}q_i^2 q_t + \frac{1}{k_t^2}q_i q_t^2 \\
&= \frac{1}{k_t^2}(q_i\mathbf{a}^2 + q_t\mathbf{a}^2 + q_i^2 q_t + q_i q_t^2) \\
&= \frac{1}{k_t^2}(q_i k_t^2 + q_t k_i^2) \\
&= q_i + q_t\frac{k_i^2}{k_t^2} = q_i + q_t\frac{\varepsilon_1}{\varepsilon_2} \\
&= \frac{\varepsilon_1}{\varepsilon_2}q_t - q_r \qquad (2.100)
\end{aligned}
$$

where $\mu_2 = \mu_1$, and

$$
\begin{aligned}
(3) \qquad & \frac{1}{k_t^2}(q_i + q_t)[(q_i - q_t)\mathbf{b} + (\mathbf{k}_i \cdot \mathbf{k}_t)\widehat{\mathbf{q}}] \\
= \; & \frac{1}{k_t^2}[(q_i^2 - q_t^2)\mathbf{b} + (q_i + q_t)(\mathbf{b}^2 + q_i q_t)\widehat{\mathbf{q}}] \\
= \; & \frac{1}{k_t^2}[(q_i^2 - q_t^2)\mathbf{b} + (q_i + q_t)\mathbf{b}^2\widehat{\mathbf{q}} + (q_i + q_t)q_i q_t\widehat{\mathbf{q}}] \\
= \; & \frac{1}{k_t^2}[q_i^2\mathbf{b} - q_t^2\mathbf{b} + q_i\mathbf{b}^2\widehat{\mathbf{q}} + q_t\mathbf{b}^2\widehat{\mathbf{q}} + q_i^2 q_t\widehat{\mathbf{q}} + q_t^2 q_i\widehat{\mathbf{q}}] \\
= \; & \frac{1}{k_t^2}[-q_t^2\mathbf{b} + q_i\mathbf{b}^2\widehat{\mathbf{q}} + q_t\mathbf{b}^2\widehat{\mathbf{q}} + q_i^2(\mathbf{b} + q_t\widehat{\mathbf{q}}) + q_t^2 q_i\widehat{\mathbf{q}}] \\
= \; & \frac{1}{k_t^2}[q_i\mathbf{b}^2\widehat{\mathbf{q}} + q_t\mathbf{b}^2\widehat{\mathbf{q}} + q_i^2(\mathbf{b} + q_t\widehat{\mathbf{q}}) - q_t^2(\mathbf{b} + q_r\widehat{\mathbf{q}})] \\
= \; & \frac{1}{k_t^2}[q_i\mathbf{b}^2\widehat{\mathbf{q}} + q_t\mathbf{b}^2\widehat{\mathbf{q}} + q_i^2\mathbf{k}_t - q_t^2\mathbf{k}_r] \\
= \; & \frac{1}{k_t^2}[q_i\mathbf{b}^2\widehat{\mathbf{q}} + q_t\mathbf{b}^2\widehat{\mathbf{q}} + q_i^2\mathbf{k}_t - q_t^2\mathbf{k}_r + \underbrace{\mathbf{b}^2\mathbf{b} - \mathbf{b}^2\mathbf{b}}_{zero}\;] \\
= \; & \frac{1}{k_t^2}[-\mathbf{b}^2(\mathbf{b} + q_r\widehat{\mathbf{q}}) + \mathbf{b}^2(\mathbf{b} + q_t\widehat{\mathbf{q}}) + q_i^2\mathbf{k}_t - q_t^2\mathbf{k}_r] \\
= \; & \frac{1}{k_t^2}[-\mathbf{b}^2\mathbf{k}_r + \mathbf{b}^2\mathbf{k}_t + q_i^2\mathbf{k}_t - q_t^2\mathbf{k}_r] \\
= \; & \frac{1}{k_t^2}[\mathbf{k}_t\mathbf{k}_i^2 - \mathbf{k}_r\mathbf{k}_t^2] = \frac{1}{k_0^2\mu_2\varepsilon_2}[\mathbf{k}_t k_0^2\mu_1\varepsilon_1 - \mathbf{k}_r\mathbf{k}_t^2] \\
= \; & \frac{\varepsilon_1}{\varepsilon_2}\mathbf{k}_t - \mathbf{k}_r
\end{aligned}
\tag{2.101}
$$

Substituting the above equation into (2.98), gives:

$$
2\cancel{q_i}\mathbf{k}_t \times \{\frac{1}{\cancel{k_t^2}}\cancel{(q_i + q_t)}[(q_i - q_t)\mathbf{b} + (\mathbf{k}_i \cdot \mathbf{k}_t)\widehat{\mathbf{q}}] \times [\widehat{\mathbf{q}} \times (\widehat{\mathbf{q}} \times \mathbf{H}_{0i})]\}
$$

$$
= \frac{1}{\cancel{k_t^2}}\cancel{(q_i + q_t)}(\mathbf{k}_i \cdot \mathbf{k}_t)\cancel{q_i}(q_i + q_t)\mathbf{H}_{0t}
$$

$$\Longrightarrow \quad 2\mathbf{k}_t \times \{\underbrace{[(q_i - q_t)\mathbf{b} + (\mathbf{k}_i \cdot \mathbf{k}_t)\widehat{\mathbf{q}}]}_{A} \times [\underbrace{\widehat{\mathbf{q}}}_{B} \times \underbrace{(\widehat{\mathbf{q}} \times \mathbf{H}_{0i})}_{C}]\}$$
$$= (\mathbf{k}_i \cdot \mathbf{k}_t)(q_i + q_t)\mathbf{H}_{0t}$$

where

$$\mathbf{A} \times (\mathbf{B} \times \mathbf{C}) = \mathbf{B}(\mathbf{A} \cdot \mathbf{C}) - \mathbf{C}(\mathbf{A} \cdot \mathbf{B})$$

$$\begin{aligned} \Longrightarrow \quad 2\mathbf{k}_t \quad & \times \quad \{\widehat{\mathbf{q}}((\widehat{\mathbf{q}} \times \mathbf{H}_{0i}) \cdot [(q_i - q_t)\mathbf{b} + (\mathbf{k}_i \cdot \mathbf{k}_t)\widehat{\mathbf{q}}]) \\ - (\widehat{\mathbf{q}} \times \mathbf{H}_{0i}) \quad & \star \quad (\widehat{\mathbf{q}} \cdot [(q_i - q_t)\mathbf{b} + (\mathbf{k}_i \cdot \mathbf{k}_t)\widehat{\mathbf{q}}])\} \\ & = \\ (\mathbf{k}_i \cdot \mathbf{k}_t)(q_i \quad & + \quad q_t)\mathbf{H}_{0t} \end{aligned}$$

$$\begin{aligned} \Longrightarrow \quad 2\mathbf{k}_t \times \{\widehat{\mathbf{q}}(-\mathbf{H}_{0i} \quad & \times \quad \widehat{\mathbf{q}} \cdot \mathbf{b}(q_i - q_t) \\ & = 0 \\ - (\mathbf{H}_{0i} \times \widehat{\mathbf{q}}) \cdot \widehat{\mathbf{q}}(\mathbf{k}_i \cdot \mathbf{k}_t)) \quad & + \quad (\mathbf{H}_{0i} \times \widehat{\mathbf{q}})(\mathbf{k}_i \cdot \mathbf{k}_t)\} \\ & = \\ (\mathbf{k}_i \cdot \mathbf{k}_t)(q_i \quad & + \quad q_t)\mathbf{H}_{0t} \end{aligned}$$

where

$$\mathbf{A} \times \mathbf{B} \cdot \mathbf{C} = \mathbf{A} \cdot \mathbf{B} \times \mathbf{C}$$

$$\begin{aligned} \Longrightarrow \quad 2\mathbf{k}_t \quad & \times \quad \{\widehat{\mathbf{q}}(\mathbf{H}_{0i} \cdot \mathbf{a})(q_i - q_t) \\ & - \quad (\widehat{\mathbf{q}} \times \mathbf{H}_{0i})(\mathbf{k}_i \cdot \mathbf{k}_t)\} \\ & = \\ (\mathbf{k}_i \cdot \mathbf{k}_t)(q_i \quad & + \quad q_t)\mathbf{H}_{0t} \end{aligned}$$

$$\begin{aligned} \Longrightarrow \quad & 2 \quad \mathbf{a}(\mathbf{H}_{0i} \cdot \mathbf{a})(q_i - q_t) \\ - \quad & 2 \quad (\mathbf{k}_t \times (\widehat{\mathbf{q}} \times \mathbf{H}_{0i}))(\mathbf{k}_i \cdot \mathbf{k}_t) \\ & = \\ (\mathbf{k}_i \cdot \mathbf{k}_t)(q_i \quad & + \quad q_t)\mathbf{H}_{0t} \end{aligned}$$

or

$$\frac{2\mathbf{a}(\mathbf{H}_{0i} \cdot \mathbf{a})(q_i - q_t) - 2(\mathbf{k}_t \times (\widehat{\mathbf{q}} \times \mathbf{H}_{0i}))(\mathbf{k}_i \cdot \mathbf{k}_t)}{(\mathbf{k}_i \cdot \mathbf{k}_t)(q_i + q_t)} = \mathbf{H}_{0t}$$

And noticing that $\mathbf{k}_t = (\mathbf{k}_i - q_i\widehat{\mathbf{q}}) + q_t\widehat{\mathbf{q}}$, then

$$\begin{aligned}
\mathbf{H}_{0t} &\\
&= \frac{2\mathbf{a}(\mathbf{H}_{0i}\cdot\mathbf{a})(q_i - q_t) - 2(\widehat{\mathbf{q}}(\mathbf{H}_{0i}\cdot\mathbf{k}_t)}{(\mathbf{k}_i\cdot\mathbf{k}_t)(q_i+q_t)} \\
&+ \frac{-\mathbf{H}_{0i}(\widehat{\mathbf{q}}\cdot\mathbf{k}_t))(\mathbf{k}_i\cdot\mathbf{k}_t)}{(\mathbf{k}_i\cdot\mathbf{k}_t)(q_i+q_t)} \\
&= \frac{2\mathbf{a}(\mathbf{H}_{0i}\cdot\mathbf{a})(q_i - q_t) - 2(\widehat{\mathbf{q}}(\mathbf{H}_{0i}\cdot(\mathbf{k}_i - q_i\widehat{\mathbf{q}} + q_t\widehat{\mathbf{q}}))}{(\mathbf{k}_i\cdot\mathbf{k}_t)(q_i+q_t)} \\
&- \frac{\mathbf{H}_{0i}(\widehat{\mathbf{q}}\cdot\mathbf{k}_t))(\mathbf{k}_i\cdot\mathbf{k}_t)}{(\mathbf{k}_i\cdot\mathbf{k}_t)(q_i+q_t)} \\
&= \frac{2\mathbf{a}(\mathbf{H}_{0i}\cdot\mathbf{a})(q_i - q_t) - 2[\widehat{\mathbf{q}}(\underbrace{\cancelto{=0}{\mathbf{H}_{0i}\cdot\mathbf{k}_i}}_{perpendicular\ to\ each\ others}}{(\mathbf{k}_i\cdot\mathbf{k}_t)(q_i+q_t)} \\
&+ \frac{-q_i\mathbf{H}_{0i}\cdot\widehat{\mathbf{q}} + q_t\mathbf{H}_{0i}\cdot\widehat{\mathbf{q}}) - \mathbf{H}_{0i}\underbrace{(\widehat{\mathbf{q}}\cdot\mathbf{k}_t)}_{=q_t}](\mathbf{k}_i\cdot\mathbf{k}_t)}{(\mathbf{k}_i\cdot\mathbf{k}_t)(q_i+q_t)}
\end{aligned}$$

Finally, we get

$$\begin{aligned}
\mathbf{H}_{0t} &= \frac{2}{q_i+q_t}[q_t\bar{\mathbf{I}} \\
&+ (q_i - q_t)(\widehat{\mathbf{q}}\widehat{\mathbf{q}} + \frac{1}{\mathbf{k}_i\cdot\mathbf{k}_t}\mathbf{aa})]\cdot\mathbf{H}_{0i}
\end{aligned} \tag{2.102}$$

Subtracting $\mathbf{H}_{0i}$ from both sides of equation (2.102) yields

$$\begin{aligned}
\mathbf{H}_{0t} - \mathbf{H}_{0i} = &\frac{2}{q_i+q_t}[q_t\bar{\mathbf{I}} + (q_i - q_t)(\widehat{\mathbf{q}}\widehat{\mathbf{q}} + \frac{1}{\mathbf{k}_i\cdot\mathbf{k}_t}\mathbf{aa})]\cdot\mathbf{H}_{0i} \\
- &\frac{\bar{\mathbf{I}}\cdot\mathbf{H}_{0i}(q_i+q_t)}{q_i+q_t}
\end{aligned}$$

$$\mathbf{H}_{0r} = \frac{q_i - q_t}{q_i + q_t}[\bar{\mathbf{I}} - 2(\widehat{\mathbf{q}}\widehat{\mathbf{q}} + \frac{1}{\mathbf{k}_i \cdot \mathbf{k}_t}\mathbf{aa})] \cdot \mathbf{H}_{0i} \tag{2.103}$$

Equations (2.102) and (2.103) are the desired results. And to show that both equations agree with equations (2.91) and (2.94) we decompose $\mathbf{H}_{0i}$ into components parallel and perpendicular to the plane of incidence:

$$\mathbf{H}_{0i} = \frac{1}{\eta_i}(\widehat{\mathbf{k}}_i \times \mathbf{E}_{0i}) = \frac{1}{\eta_i}[A_\perp(\widehat{\mathbf{k}_i} \times \mathbf{a}) - A_\parallel \mathbf{a}]$$

And then substitute the result into equations (2.102) and (2.103)

$$\mathbf{H}_{0r} = \frac{q_i - q_t}{q_i + q_t}[\bar{\mathbf{I}} - 2(\widehat{\mathbf{q}}\widehat{\mathbf{q}} + \frac{1}{\mathbf{k}_i \cdot \mathbf{k}_t}\mathbf{aa})] \cdot \{\frac{1}{\eta_i}[A_\perp(\widehat{\mathbf{k}_i} \times \mathbf{a}) - A_\parallel \mathbf{a}]\}$$

where

$$\begin{aligned}
(1) &\quad \mathbf{a} \cdot (\widehat{\mathbf{k}_i} \times \mathbf{a}) = 0 \\
(2) &\quad \widehat{\mathbf{q}} \cdot (\widehat{\mathbf{k}_i} \times \mathbf{a}) = -(\widehat{\mathbf{q}} \times \mathbf{a}) \cdot \widehat{\mathbf{k}_i} = -\mathbf{b} \cdot \widehat{\mathbf{k}_i} = -\frac{\mathbf{a}^2}{|k_i|} \\
(3) &\quad \frac{2\mathbf{a}^2}{\mathbf{k}_i \cdot \mathbf{k}_t} - 1 = \frac{2\mathbf{a}^2 - \mathbf{a}^2 - q_i q_t}{\mathbf{k}_i \cdot \mathbf{k}_t} = \frac{\mathbf{a}^2 + q_r q_t}{\mathbf{k}_i \cdot \mathbf{k}_t} \\
= &\quad \underbrace{(\frac{\mathbf{k}_r \cdot \mathbf{k}_t}{\mathbf{k}_i \cdot \mathbf{k}_t})}_{|\mathbf{k}_t| = |\mathbf{k}_r|} = (\frac{\widehat{\mathbf{k}}_r \cdot \widehat{\mathbf{k}}_t}{\widehat{\mathbf{k}}_i \cdot \widehat{\mathbf{k}}_t}) \\
(4) &\quad q_t + \frac{1}{\mathbf{k}_i \cdot \mathbf{k}_t}(q_i - q_t)\mathbf{a}^2 = \frac{\cancel{q_t \mathbf{a}^2} + q_t^2 q_i + q_i \mathbf{a}^2 - \cancel{q_t \mathbf{a}^2}}{\mathbf{k}_i \cdot \mathbf{k}_t} \\
= &\quad \frac{q_i k_t}{k_i(\widehat{\mathbf{k}}_i \cdot \widehat{\mathbf{k}}_t)}
\end{aligned}$$

Then

$$
\begin{aligned}
\mathbf{H}_{0r} &= (\frac{1}{\eta_i})(\frac{q_i - q_t}{q_i + q_t})[A_\perp(\widehat{\mathbf{k}_i} \times \mathbf{a}) - A_\parallel \mathbf{a} \\
&\quad - 2\widehat{\mathbf{q}}\widehat{\mathbf{q}} \cdot A_\perp(\widehat{\mathbf{k}_i} \times \mathbf{a}) + 2\frac{1}{\mathbf{k}_i \cdot \mathbf{k}_t}\mathbf{aa} \cdot A_\parallel \mathbf{a}] \\
&= (\frac{1}{\eta_i})(\frac{q_i - q_t}{q_i + q_t})[A_\perp(\widehat{\mathbf{k}_i} \times \mathbf{a}) - A_\parallel \mathbf{a} \\
&\quad - 2A_\perp\widehat{\mathbf{q}}\widehat{\mathbf{q}} \cdot (\widehat{\mathbf{k}_i} \times \mathbf{a}) + 2\frac{A_\parallel}{\mathbf{k}_i \cdot \mathbf{k}_t}\mathbf{aa} \cdot \mathbf{a}] \\
&= (\frac{1}{\eta_i})(\frac{q_i - q_t}{q_i + q_t})[A_\perp(\widehat{\mathbf{k}_i} \times \mathbf{a}) - A_\parallel \mathbf{a} \\
&\quad - 2A_\perp\widehat{\mathbf{q}}(\frac{-\mathbf{a}^2}{|k_i|}) + 2\frac{A_\parallel}{\mathbf{k}_i \cdot \mathbf{k}_t}\mathbf{a}^2\mathbf{a}] \\
&= (\frac{1}{\eta_i})(\frac{q_i - q_t}{q_i + q_t})[A_\perp(\widehat{\mathbf{k}_i} \times \mathbf{a}) \\
&\quad + 2A_\perp\widehat{\mathbf{q}}(\frac{\mathbf{a}^2}{|k_i|}) + A_\parallel \mathbf{a}\frac{\widehat{\mathbf{k}}_r \cdot \widehat{\mathbf{k}}_t}{\widehat{\mathbf{k}}_i \cdot \widehat{\mathbf{k}}_t}]
\end{aligned}
\tag{2.104}
$$

And from equation (2.102)

$$\begin{aligned}
\mathbf{H}_{0t} &= \frac{2}{q_i+q_t}[q_t\bar{\mathbf{I}} + (q_i - q_t)(\widehat{\mathbf{q}}\widehat{\mathbf{q}} + \frac{1}{\mathbf{k}_i\cdot\mathbf{k}_t}\mathbf{aa})] \\
&\cdot (\frac{1}{\eta_i}[A_\perp(\widehat{\mathbf{k}_i}\times\mathbf{a}) - A_\parallel \mathbf{a}]) \\
&= (\frac{1}{\eta_i})(\frac{2}{q_i+q_t})[q_t\bar{\mathbf{I}} + (q_i - q_t)(\widehat{\mathbf{q}}\widehat{\mathbf{q}} + \frac{1}{\mathbf{k}_i\cdot\mathbf{k}_t}\mathbf{aa})] \\
&\cdot \{A_\perp(\widehat{\mathbf{k}_i}\times\mathbf{a}) - A_\parallel \mathbf{a}\} \\
&= (\frac{1}{\eta_i})(\frac{2}{q_i+q_t})[q_t A_\perp(\widehat{\mathbf{k}_i}\times\mathbf{a}) - q_t A_\parallel \mathbf{a} \\
&+ (q_i - q_t)(\widehat{\mathbf{q}}\widehat{\mathbf{q}}\cdot A_\perp(\widehat{\mathbf{k}_i}\times\mathbf{a}) - \frac{1}{\mathbf{k}_i\cdot\mathbf{k}_t}\mathbf{aa}\cdot A_\parallel \mathbf{a})] \\
&= (\frac{1}{\eta_i})(\frac{2}{q_i+q_t})[q_t A_\perp(\widehat{\mathbf{k}_i}\times\mathbf{a}) - q_t A_\parallel \mathbf{a} \\
&+ (q_i - q_t)A_\perp\widehat{\mathbf{q}}(\frac{-\mathbf{a}^2}{|k_i|}) - (q_i - q_t)\frac{A_\parallel}{\mathbf{k}_i\cdot\mathbf{k}_t}\mathbf{a}^2\mathbf{a}] \\
&= (\frac{1}{\eta_i})(\frac{2}{q_i+q_t})[q_t A_\perp(\widehat{\mathbf{k}_i}\times\mathbf{a}) - A_\parallel \mathbf{a}(\frac{q_i k_t}{k_i(\widehat{\mathbf{k}}_i\cdot\widehat{\mathbf{k}}_t)}) \\
&- (q_i - q_t)A_\perp\widehat{\mathbf{q}}(\frac{\mathbf{a}^2}{|k_i|})] \qquad (2.105)
\end{aligned}$$

Regarding equation (2.95), the electric field is

$$\mathbf{E}_{0r} = \frac{q_i - q_t}{q_i+q_t}[A_\perp\mathbf{a} + \frac{\widehat{\mathbf{k}}_r\cdot\widehat{\mathbf{k}}_t}{\widehat{\mathbf{k}}_i\cdot\widehat{\mathbf{k}}_t}A_\parallel(\widehat{\mathbf{k}}_r\times\mathbf{a})] \qquad (2.106)$$

$$\mathbf{E}_{0t} = \frac{2q_i}{q_i+q_t}[A_\perp\mathbf{a} + \frac{1}{\widehat{\mathbf{k}}_i\cdot\widehat{\mathbf{k}}_t}A_\parallel(\widehat{\mathbf{k}}_t\times\mathbf{a})] \qquad (2.107)$$

These following equations have been used in finding (2.106) and (2.107)

(1) $\widehat{\mathbf{k}}_t \times (\widehat{\mathbf{k}}_i \times \mathbf{a}) = \widehat{\mathbf{k}}_i \overbrace{(\mathbf{a} \cdot \widehat{\mathbf{k}}_t)}^{=0} - \mathbf{a}(\widehat{\mathbf{k}}_t \cdot \widehat{\mathbf{k}}_i)$

(2) $\widehat{\mathbf{k}}_t \times \widehat{\mathbf{q}} = \dfrac{\mathbf{a}}{|k_t|}$

(4) $k_t\eta_t = \dfrac{w\sqrt{\mu_2\varepsilon_2}}{c} \dfrac{\sqrt{\mu_2\mu_0}}{\sqrt{\varepsilon_2\varepsilon_0}} = \dfrac{\omega\mu_2}{c}\sqrt{\dfrac{\mu_0}{\varepsilon_0}}$

$$= w\mu_2\sqrt{\frac{\mu_0}{\varepsilon_0}}\sqrt{\varepsilon_0\mu_0} = \omega\mu_0\mu_2$$

$$\Longrightarrow \frac{k_t\eta_t}{k_i\eta_i} = \frac{\mu_2}{\mu_1} = 1$$

(5) $\dfrac{2\mathbf{a}^2}{k_i^2} - (\widehat{\mathbf{k}}_t \cdot \widehat{\mathbf{k}}_r) = 1$

$$\Longrightarrow 2\mathbf{a}^2 - \mathbf{k}_t \cdot \mathbf{k}_r = k_i^2$$

$$\Longrightarrow 2\mathbf{a}^2 - \mathbf{a}^2 - q_i q_r = k_i^2 = k_i^2 - \mathbf{a}^2 = q_i q_r$$

$$\Longrightarrow q_i = -q_r$$

(6) $q_t(\widehat{\mathbf{k}}_t \cdot \widehat{\mathbf{k}}_i) + \dfrac{(q_i - q_t)\mathbf{a}^2}{k_i k_t} = \dfrac{q_i k_t}{k_i}$

2.2.2. Fresnel Equations: Normal Incidence

When a uniform plane wave is normally incident upon an interface of two media, the method of decomposing the amplitude vectors into components perpendicular and parallel!to plane of incidence to the plane of incidence fails since $\mathbf{k}_i || \widehat{\mathbf{q}}$ means

$$\mathbf{a} = \mathbf{k}_i \times \widehat{\mathbf{q}} = \mathbf{k}_r \times \widehat{\mathbf{q}} = \mathbf{k}_t \times \widehat{\mathbf{q}} = \mathbf{0} \qquad (2.108)$$

and the concept of plane of incidence loses its meaning.

According to equation (2.108), we may write $\widehat{\mathbf{k}}_i = \widehat{\mathbf{k}}_t = -\widehat{\mathbf{k}}_r = \widehat{\mathbf{q}}$ as we observe on Figure 2.3 below.

$$\begin{aligned} \mathbf{k}_i &= k_i\widehat{\mathbf{k}}_i = k_i\widehat{\mathbf{q}} \\ \mathbf{k}_r &= k_r\widehat{\mathbf{k}}_r = -k_i\widehat{\mathbf{q}} \\ \mathbf{k}_t &= k_t\widehat{\mathbf{k}}_t = k_t\widehat{\mathbf{q}} \end{aligned} \qquad (2.109)$$

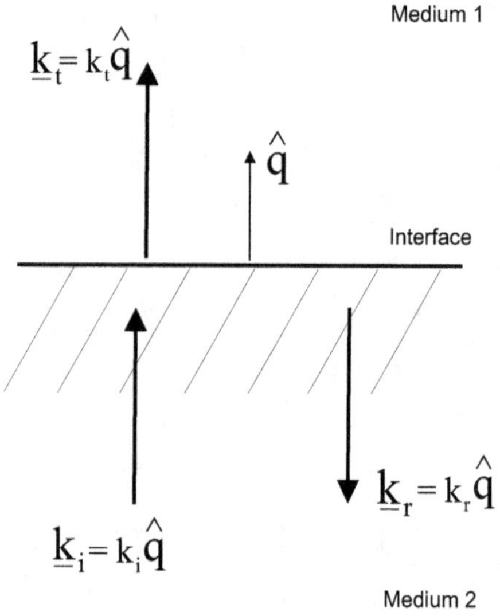

Figure 2.3.: Normal Incidence of a Uniform Plane Wave on a Lossless Medium.

Therefore, from Maxwell's equation $\mathbf{k} \cdot \mathbf{E}_0 = 0$,

$$\widehat{\mathbf{q}} \cdot \mathbf{E}_{0i} = \widehat{\mathbf{q}} \cdot \mathbf{E}_{0r} = \widehat{\mathbf{q}} \cdot \mathbf{E}_{0t} = 0 \tag{2.110}$$

at every point in space and on the interface. From equation (2.110), we write

$$(\mathbf{E}_{0i} + \mathbf{E}_{0r} - \mathbf{E}_{0t}) \cdot \widehat{\mathbf{q}} = 0 \tag{2.111}$$

which together with the boundary condition (2.37) implies that the vector $\mathbf{E}_{0i} + \mathbf{E}_{0r} - \mathbf{E}_{0t}$ is simultaneously parallel and perpendicular to the vector $\widehat{\mathbf{q}}$. This is only possible if the vector itself is a zero vector, i.e.,

$$\mathbf{E}_{0i} + \mathbf{E}_{0r} - \mathbf{E}_{0t} = \mathbf{0} \tag{2.112}$$

On the other hand, substituting $\mathbf{H}_0 = (1/\eta)\widehat{\mathbf{k}} \times \mathbf{E}_0$ into equation (2.38) and expanding the vector triple product, after using the condition (2.110), we obtain

$$\begin{aligned} ((1/\eta_i)\widehat{\mathbf{k}_i} \times \mathbf{E}_{0i} + (1/\eta_i)\widehat{\mathbf{k}_r} &\times \mathbf{E}_{0r} \\ - (1/\eta_t)\widehat{\mathbf{k}_t} \times \mathbf{E}_{0t}) \times \widehat{\mathbf{q}} &= \mathbf{0} \\ \eta_t \mathbf{E}_{0i} - \eta_t \mathbf{E}_{0r} - \eta_i \mathbf{E}_{0t} &= \mathbf{0} \end{aligned} \tag{2.113}$$

Solving for equations (2.112) and (2.113), we find

$$\mathbf{E}_{0r} = \frac{\eta_t - \eta_i}{\eta_t + \eta_i}\mathbf{E}_{0i} \tag{2.114}$$

$$\mathbf{E}_{0t} = \frac{2\eta_t}{\eta_t + \eta_i}\mathbf{E}_{0i} \tag{2.115}$$

And by taking the cross product of the last two equations with $(1/\eta_i)\widehat{\mathbf{k}}_r\times$ and $(1/\eta_t)\widehat{\mathbf{k}}_t\times$ respectively, we have

$$\mathbf{H}_{0r} = \frac{\eta_i - \eta_t}{\eta_t + \eta_i}\mathbf{H}_{0i} \tag{2.116}$$

$$\begin{aligned} \mathbf{H}_{0t} &= \frac{2\eta_t}{\eta_t + \eta_i}(1/\eta_t)\widehat{\mathbf{k}}_t \times \mathbf{E}_{0i} \\ &= \frac{2\eta_t}{\eta_t + \eta_i}(\frac{1}{\eta_t})\widehat{\mathbf{k}_i} \times \mathbf{E}_{0i}(\frac{\eta_i}{\eta_i}) \\ &= \frac{2\eta_i}{\eta_t + \eta_i}\mathbf{H}_{0i} \end{aligned} \tag{2.117}$$

where, $\widehat{\mathbf{k}}_r = -\widehat{\mathbf{k}}_i$ and $\widehat{\mathbf{k}}_t = \widehat{\mathbf{k}}_i$.

In the above derivation, we have used the general boundary conditions (2.37), (2.38) and the normal incidence condition (2.109). Therefore, equations (2.114) to (2.117) are valid for media with different permeabilities as well as losses.

For media having the same permeability, equations (2.116) and (2.117) reduce to

$$\mathbf{H}_{0r} = \frac{\sqrt{\frac{\mu_1}{\varepsilon_1}} - \sqrt{\frac{\mu_2}{\varepsilon_2}}}{\sqrt{\frac{\mu_1}{\varepsilon_1}} + \sqrt{\frac{\mu_2}{\varepsilon_2}}}\mathbf{H}_{0i} = \frac{q_t - q_i}{q_t + q_i}\mathbf{H}_{0i} \tag{2.118}$$

$$\mathbf{H}_{0t} = \frac{2\sqrt{\frac{\mu_1}{\varepsilon_1}}}{\sqrt{\frac{\mu_1}{\varepsilon_1}} + \sqrt{\frac{\mu_2}{\varepsilon_2}}}\mathbf{H}_{0i} = \frac{2q_t}{q_t + q_i}\mathbf{H}_{0i} \tag{2.119}$$

where

$$q_i = \sqrt{k_i^2 - \mathbf{a}^2} = \sqrt{k_0^2\varepsilon_1 - (k_i \sin\theta_i)^2}$$
$$q_t = \sqrt{k_t^2 - \mathbf{a}^2} = \sqrt{k_0^2\varepsilon_2 - (k_t \sin\theta_t)^2}$$

but $\sin\theta_t = \sin\theta_i = \sin(0) = 0$

$$q_i = \sqrt{k_0^2\varepsilon_1}$$
$$q_t = \sqrt{k_0^2\varepsilon_2}$$

2.3. Reflection and Transmission of Plane Waves Propagating in Lossless, Nonmagnetic and Unbounded Isotropic Media through a Plate of Lossless, Nonmagnetic and Bounded Homogeneous Isotropic Media

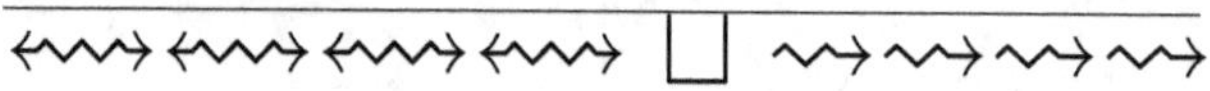

For a uniform plane wave propagating through an isotropic plate perpendicular to its interface as shown in Figure 2.4 below, the overall summation of the reflected waves on the outside surface is $\mathbf{E}_r$ and on the inside surface of the plate is $\mathbf{E}_{d_2}$.

The vectors[2] are

$$\mathbf{E}_i = E_i e^{-jk_0 z}\underline{\mathbf{e}}_x \quad (2.120)$$
$$\mathbf{E}_r = E_r e^{jk_0 z}\underline{\mathbf{e}}_x \quad (2.121)$$
$$\mathbf{E}_{d_1} = E_{d_1} e^{-jk_d z}\underline{\mathbf{e}}_x \quad (2.122)$$
$$\mathbf{E}_{d_2} = E_{d_2} e^{jk_d z}\underline{\mathbf{e}}_x \quad (2.123)$$
$$\mathbf{E}_T = E_T e^{-jk_0 z}\underline{\mathbf{e}}_x \quad (2.124)$$

[2] Some books use the other convention $\mathbf{E}(\mathbf{Z}, \omega)$ instead of $\mathbf{E}$ (which is used in this book) to show the vector in Frequency Domain.

And

$$\mathbf{H}_i = -\frac{E_i}{\eta_0} e^{-jk_0 z} \underline{\mathbf{e}}_y \tag{2.125}$$

$$\mathbf{H}_r = \frac{E_r}{\eta_0} e^{jk_0 z} \underline{\mathbf{e}}_y \tag{2.126}$$

$$\mathbf{H}_{d_1} = -\frac{E_{d_1}}{\eta_d} e^{-jk_d z} \underline{\mathbf{e}}_y \tag{2.127}$$

$$\mathbf{H}_{d_2} = \frac{E_{d_2}}{\eta_d} e^{jk_d z} \underline{\mathbf{e}}_y \tag{2.128}$$

$$\mathbf{H}_T = -\frac{E_T}{\eta_0} e^{-jk_0 z} \underline{\mathbf{e}}_y \tag{2.129}$$

1) First surface boundary condition when z=d

$$\begin{aligned} (A) \quad & \mathbf{n} \times (\mathbf{E}_i(z=d) + \mathbf{E}_r(z=d)) \\ & = \mathbf{n} \times (\mathbf{E}_{d_1}(z=d) + \mathbf{E}_{d_2}(z=d)) \\ (B) \quad & \mathbf{n} \times (\mathbf{H}_i(z=d) + \mathbf{H}_r(z=d)) \\ & = \mathbf{n} \times (\mathbf{H}_{d_1}(z=d) + \mathbf{H}_{d_2}(z=d)) \end{aligned}$$

We have[3]

$$\begin{aligned} (A) \quad & \underline{\mathbf{e}}_z \times (E_i e^{-jk_0 d} \underline{\mathbf{e}}_x + E_r e^{jk_0 d} \underline{\mathbf{e}}_x) \\ & = \underline{\mathbf{e}}_z \times (E_{d_1} e^{-jk_d d} \underline{\mathbf{e}}_x + E_{d_2} e^{jk_d d} \underline{\mathbf{e}}_x) \\ (B) \quad & \underline{\mathbf{e}}_z \times (-\frac{E_i}{\eta_0} e^{-jk_0 d} \underline{\mathbf{e}}_y + \frac{E_r}{\eta_0} e^{jk_0 d} \underline{\mathbf{e}}_y) \\ & = \underline{\mathbf{e}}_z \times (-\frac{E_{d_1}}{\eta_d} e^{-jk_d d} \underline{\mathbf{e}}_y + \frac{E_{d_2}}{\eta_d} e^{jk_d d} \underline{\mathbf{e}}_y) \end{aligned}$$

[3] $\mathbf{n} = \underline{\mathbf{e}}_z$

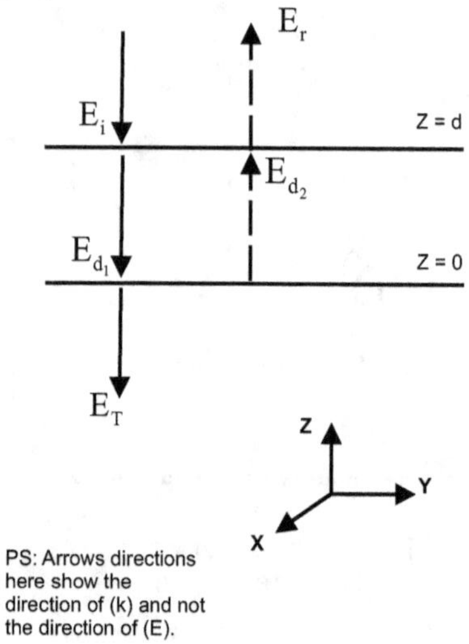

Figure 2.4.: Uniform Plane Wave Propagating through an Isotropic Plate.

And by removing the unit vectors

$$(A) \qquad E_i e^{-jk_0 d} + E_r e^{jk_0 d} = E_{d_1} e^{-jk_d d} + E_{d_2} e^{jk_d d}$$

$$(B) \qquad -\frac{E_i}{\eta_0} e^{-jk_0 d} + \frac{E_r}{\eta_0} e^{jk_0 d} = -\frac{E_{d_1}}{\eta_d} e^{-jk_d d} + \frac{E_{d_2}}{\eta_d} e^{jk_d d}$$

Then combining the exponential terms in (A) with each others

$$E_i e^{-jd(k_0+k_d)} + E_r e^{jd(k_0-k_d)} - E_{d_1} e^{-2jk_d d} = E_{d_2} \qquad (2.130)$$

And substituting the latter equation into (B), we get

$$(A) \quad E_i e^{-jd(k_0+k_d)} + E_r e^{jd(k_0-k_d)} - E_{d_1} e^{-2jk_d d} = E_{d_2}$$

$$\begin{aligned}(B) \qquad & -\frac{E_i}{\eta_0} e^{-jk_0 d} + \frac{E_r}{\eta_0} e^{jk_0 d} \\ = & -\frac{E_{d_1}}{\eta_d} e^{-jk_d d} + \frac{E_i e^{-jd(k_0+k_d)} + E_r e^{jd(k_0-k_d)}}{\eta_d} e^{jk_d d} \\ & - \frac{E_{d_1} e^{-2jk_d d}}{\eta_d} e^{jk_d d}\end{aligned}$$

$\Longrightarrow$

$$\begin{aligned}(A) \qquad & E_i e^{-jd(k_0+k_d)} + E_r e^{jd(k_0-k_d)} - E_{d_1} e^{-2jk_d d} = E_{d_2} \\ (B) \qquad & -\frac{E_i}{\eta_0} e^{-jk_0 d} + \frac{E_r}{\eta_0} e^{jk_0 d} \\ = & -\frac{E_{d_1}}{\eta_d} e^{-jk_d d} + \frac{E_i}{\eta_d} e^{-jd(k_0)} + \frac{E_r}{\eta_d} e^{jd(k_0)} - \frac{E_{d_1}}{\eta_d} e^{-jk_d d}\end{aligned}$$

And finally we combine the terms in (B)

$$\begin{aligned}
\Longrightarrow \quad & - \frac{E_i}{\eta_0}e^{-jk_0 d} - \frac{E_i}{\eta_d}e^{-jdk_0} + \frac{E_r}{\eta_0}e^{jk_0 d} - \frac{E_r}{\eta_d}e^{jdk_0} \\
= \quad & - 2\frac{E_{d_1}}{\eta_d}e^{-jk_d d}
\end{aligned}$$

$$\begin{aligned}
& E_i\left(\frac{e^{-jk_0 d}}{\eta_0} + \frac{e^{-jdk_0}}{\eta_d}\right) + E_r\left(\frac{-e^{jk_0 d}}{\eta_0} + \frac{e^{jdk_0}}{\eta_d}\right) \\
= \quad & 2\frac{E_{d_1}}{\eta_d}e^{-jk_d d}
\end{aligned}$$

$$\begin{aligned}
& \left(\frac{e^{-jk_0 d}}{\eta_0} + \frac{e^{-jdk_0}}{\eta_d}\right) + \left(\frac{E_r}{E_i}\right)\left(\frac{-e^{jk_0 d}}{\eta_0} + \frac{e^{jdk_0}}{\eta_d}\right) \\
= \quad & 2\frac{e^{-jk_d d}}{\eta_d}\frac{E_{d_1}}{E_i}
\end{aligned}$$

$$\begin{aligned}
& \frac{(\eta_d e^{-jk_0 d} + \eta_0 e^{-jdk_0}) + (\frac{E_r}{E_i})(-\eta_d e^{jk_0 d} + \eta_0 e^{jdk_0})}{2e^{-jk_d d}\eta_0} \\
= \quad & \frac{E_{d_1}}{E_i}
\end{aligned} \tag{2.131}$$

2) Second surface boundary condition when z=0

$$\begin{aligned}
(A) \quad & \underline{\mathbf{e}}_z \times (E_{d_1}\cancelto{=1}{e^{-jk_d z}}\underline{\mathbf{e}}_x + E_{d_2}\cancelto{=1}{e^{jk_d z}}\underline{\mathbf{e}}_x) = \\
& \underline{\mathbf{e}}_z \times (E_T\cancelto{=1}{e^{-jk_0 z}}\underline{\mathbf{e}}_x)
\end{aligned}$$

$$\Longrightarrow \quad E_{d_1} + E_{d_2} = E_T \tag{2.132}$$

$$
\begin{aligned}
(B) \qquad & \underline{\mathbf{e}}_z \times \left(\frac{E_{d_1}}{\eta_d}\cancelto{=1}{e^{-jk_d z}}\,\underline{\mathbf{e}}_y - \frac{E_{d_2}}{\eta_d}\cancelto{=1}{e^{jk_d z}}\,\underline{\mathbf{e}}_y\right) = \\
& \underline{\mathbf{e}}_z \times \left(\frac{E_T}{\eta_0}\cancelto{=1}{e^{-jk_0 z}}\,\underline{\mathbf{e}}_y\right) \\
\Longrightarrow \quad & \frac{E_{d_1} - E_{d_2}}{\eta_d} = \frac{E_T}{\eta_0} \qquad (2.133)
\end{aligned}
$$

Substituting (2.132) in (2.133)

$$
\begin{aligned}
\frac{\eta_0}{\eta_d}(E_{d_1} - E_{d_2}) &= E_{d_1} + E_{d_2} \\
(\frac{\eta_0}{\eta_d} - 1)E_{d_1} &= (\frac{\eta_0}{\eta_d} + 1)E_{d_2} \\
E_{d_2} &= \frac{\eta_0 - \eta_d}{\eta_0 + \eta_d}E_{d_1} \qquad (2.134)
\end{aligned}
$$

Substituting (2.134) in (2.130)

$$
\begin{aligned}
E_i e^{-jd(k_0+k_d)} + E_r e^{jd(k_0-k_d)} &- E_{d_1} e^{-2jk_d d} \\
&= \frac{\eta_0 - \eta_d}{\eta_0 + \eta_d}E_{d_1} \\
E_i e^{-jd(k_0+k_d)} + E_r e^{jd(k_0-k_d)} &- E_{d_1}(e^{-2jk_d d} + \frac{\eta_0 - \eta_d}{\eta_0 + \eta_d}) \\
&= 0 \\
e^{-jdk_0} + (\frac{E_r}{E_i})e^{jdk_0} - \frac{E_{d_1}}{E_i}e^{-jk_d d}(1 &+ \frac{\eta_0 - \eta_d}{\eta_0 + \eta_d}e^{2jk_d d}) \\
&= 0 \\
\frac{e^{-jdk_0} + (\frac{E_r}{E_i})e^{jdk_0}}{e^{-jk_d d}(1 + \frac{\eta_0-\eta_d}{\eta_0+\eta_d}e^{2jk_d d})} &= \frac{E_{d_1}}{E_i} \qquad (2.135)
\end{aligned}
$$

then, (2.131) = (2.135)

$$\frac{e^{-jdk_0} + (\frac{E_r}{E_i})e^{jdk_0}}{\cancel{e^{-jk_d d}}(1 + \frac{\eta_0 - \eta_d}{\eta_0 + \eta_d} e^{2jk_d d})} = \frac{(\eta_d e^{-jk_0 d} + \eta_0 e^{-jdk_0}) + (\frac{E_r}{E_i})(-\eta_d e^{jk_0 d} + \eta_0 e^{jdk_0})}{2\cancel{e^{-jk_d d}}\eta_0}$$

$$2\eta_0(e^{-jdk_0} + (\frac{E_r}{E_i})e^{jdk_0}) = (1 + \frac{\eta_0 - \eta_d}{\eta_0 + \eta_d} e^{2jk_d d})\{e^{-jdk_0}(\eta_d + \eta_0) + e^{jdk_0}(\frac{E_r}{E_i})(-\eta_d + \eta_0)\}$$

$$\begin{aligned} 2\eta_0 e^{-jdk_0} + 2\eta_0(\frac{E_r}{E_i})e^{jdk_0} &= e^{-jdk_0}(\eta_d + \eta_0)(1 + \frac{\eta_0 - \eta_d}{\eta_0 + \eta_d} e^{2jk_d d}) \\ &\quad + e^{jdk_0}(\frac{E_r}{E_i})(-\eta_d + \eta_0)(1 + \frac{\eta_0 - \eta_d}{\eta_0 + \eta_d} e^{2jk_d d}) \end{aligned}$$

getting

$$\begin{aligned} (\frac{E_r}{E_i}) &= \frac{e^{-jdk_0}(\eta_d + \eta_0)(1 + \frac{\eta_0 - \eta_d}{\eta_0 + \eta_d} e^{2jk_d d}) - 2\eta_0 e^{-jdk_0}}{2\eta_0 e^{jdk_0} - e^{jdk_0}(-\eta_d + \eta_0)(1 + \frac{\eta_0 - \eta_d}{\eta_0 + \eta_d} e^{2jk_d d})} \\ &= e^{-2jdk_0}\frac{(\eta_d + \eta_0)(1 + \frac{\eta_0 - \eta_d}{\eta_0 + \eta_d} e^{2jk_d d}) - 2\eta_0}{2\eta_0 - (-\eta_d + \eta_0)(1 + \frac{\eta_0 - \eta_d}{\eta_0 + \eta_d} e^{2jk_d d})} \end{aligned} \tag{2.136}$$

where the reflection coefficient Γ is defined as the reflected electric field on the first surface over the incident electric field on the first surface

$$\Gamma = \frac{E_r e^{jk_0 d}}{E_i e^{-jk_0 d}} = \frac{E_r}{E_i} e^{2jk_0 d} \tag{2.137}$$

$$= \frac{(\eta_d + \eta_0)(1 + \frac{\eta_0 - \eta_d}{\eta_0 + \eta_d} e^{2jk_d d}) - 2\eta_0}{2\eta_0 - (-\eta_d + \eta_0)(1 + \frac{\eta_0 - \eta_d}{\eta_0 + \eta_d} e^{2jk_d d})} \tag{2.138}$$

From (2.132) and (2.133), we have

$$\begin{aligned} \frac{1}{\eta_d}(2E_{d_1} - E_T) &= \frac{E_T}{\eta_0} \\ E_T(\frac{1}{\eta_d} + \frac{1}{\eta_0}) &= \frac{2E_{d_1}}{\eta_d} \end{aligned} \tag{2.139}$$

Dividing both sides of equation (2.139) over E_i gives

$$\begin{aligned} \frac{E_T}{E_i} &= \frac{2}{(\frac{1}{\eta_d} + \frac{1}{\eta_0})} \frac{1}{\eta_d} \frac{E_{d_1}}{E_i} \\ &= \frac{2}{(1 + \frac{\eta_d}{\eta_0})} \frac{E_{d_1}}{E_i} \\ &= \frac{2}{(1 + \frac{\eta_d}{\eta_0})} * \\ &\frac{(\eta_d e^{-jk_0 d} + \eta_0 e^{-jdk_0}) + (\frac{E_T}{E_i})(-\eta_d e^{jk_0 d} + \eta_0 e^{jdk_0})}{2e^{-jk_d d}\eta_0} \\ &= \frac{2}{(1 + \frac{\eta_d}{\eta_0})} * \\ &\frac{e^{-jk_0 d}(\eta_d + \eta_0) + (\Gamma e^{-2jk_0 d})(-\eta_d e^{jk_0 d} + \eta_0 e^{jdk_0})}{2e^{-jk_d d}\eta_0} \end{aligned} \tag{2.140}$$

The transmission coefficient T is defined as the transmitted electric field out of the second surface over the incident electric field on the first surface

$$T = \frac{E_T e^{-jk_0 z}|_{z=0}}{E_i e^{-jk_0 z}|_{z=d}} = \frac{E_T}{E_i} e^{jk_0 d} \tag{2.141}$$

$$= (\frac{2}{1 + \frac{\eta_d}{\eta_0}}) \frac{(\eta_d + \eta_0) + (\Gamma)(-\eta_d + \eta_0)}{2\eta_0 e^{-jk_d d}} \tag{2.142}$$

Applying a propagating RC_2 pulse through the isotropic plate and using the earlier analytic results of reflection and transmission coefficients in calculating the reflected and transmitted electric fields (in time and frequency domains) will show us the relation between the different parameters, i.e. thickness of the plate d; frequency of RC_2 as f; and the permittivity of the medium ε.

In Figure 2.5 below we see the following signals: **Inc**: Incident raised cosine pulse on the first interface of the plate; $\mathbf{T_1}$: First transmitted raised cosine pulse coming out of the plate; $\mathbf{T_2}$: Second transmitted raised cosine pulse coming out of the plate; $\mathbf{T_3}$: Third transmitted raised cosine pulse coming out of the plate. And $\mathbf{R_1}$: First reflected raised cosine pulse off the first interface of the plate from the incident wave; $\mathbf{R_2}$: Second reflected raised cosine pulse off the second interface of the plate; $\mathbf{R_3}$: Third reflected raised cosine pulse off the second interface of the plate.

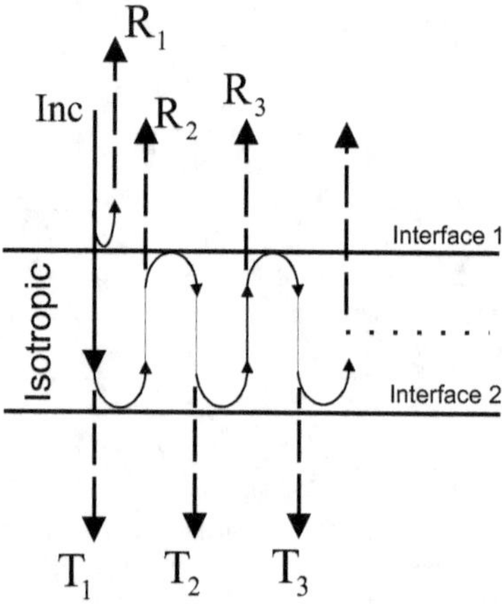

Figure 2.5.: Normal Incidence of the RC_2 Pulse on an Isotropic Plate with the directions of the Reflected and Transmitted Waves.

The frequency of the pulse which I am applying equals to 10^9 Hertz, and the plate's thickness equals to 0.3 meter; with an isotropic medium having a relative permittivity of 1 for the first medium and an isotropic plate having a relative permittivity of 5.

By having the relation $v_p = c_0/\sqrt{\varepsilon_r}$, we can find that the phase velocity inside the isotropic plate equals to $1.3407 \star 10^8$ m/sec. Dividing the plate thickness by the phase velocity we find the time needed for the raised cosine pulse to reach the second interface of the plate from the first one which equals to $0.22376 \star 10^{-8}$ seconds.

This means that each bouncing of the reflected raised cosine pulse off any of the two interfaces inside the plate needs an amount of time of $0.22376 \star 10^{-8}$ seconds for each reflection from one interface to the other. This corresponds with distances between the resulted signals on points of 0, $0.22376 \star 10^{-8}$, $0.4475 \star 10^{-8}$, $0.6713 \star 10^{-8}$ and $0.8950 \star 10^{-8}$ seconds, and so on.

And in Figure 2.6 below, we see the relation in time between the raised cosine incident, reflected and transmitted pulses on the external and internal interfaces of the plate.

The effect of changing the relative permittivity of the isotropic plate into 15, can be seen in Figure 2.7 below.

Note that $v_p = c_0/\sqrt{\varepsilon_{new}}$ has reached $0.77 * 10^8$ m/sec; and by using this value we can calculate the needed time for the signal to travel from an interface to the other, i.e. $0.3876 * 10^{-8}$ seconds.

Another thing to observe in the figures above is that R_1 has been flipped over in regard to the other reflected signals, and that shows that the isotropic plate has a higher relative permittivity value in comparison to that of the first medium's whence the wave is propagating, as shown in equation (2.114).

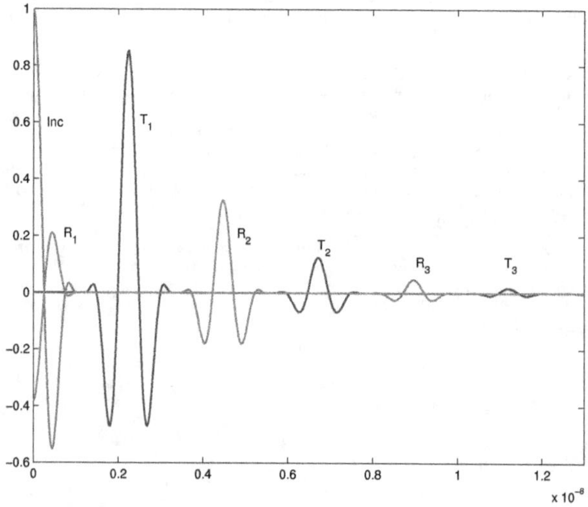

Figure 2.6.: Time Domain Representation of the Transmitted, Reflected and Incident RC_2 Pulses through an Isotropic Plate with a Relative Permittivity of 5.

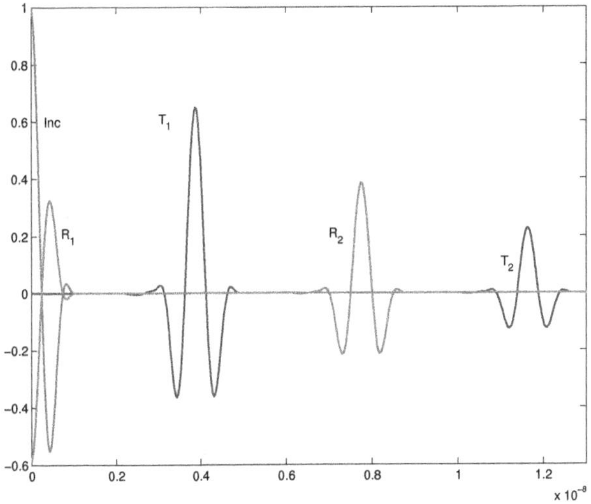

Figure 2.7.: Time Domain Representation of the Transmitted, Reflected and Incident RC_2 Pulses through an Isotropic Plate with a Relative Permittivity of 15.

Part III.

ANISOTROPIC CASE

3

Plane Waves in Crystals

3.1. Plane Waves in Lossless, Nonmagnetic and Homogeneous Crystals

The phase velocity in crystals depends on the direction of wave propagation. And to each direction of wave propagation correspond two linearly polarized waves traveling at different phase velocities. This phenomenon is known as birefringence or double refraction.

3.1.1. Properties of the Dielectric Tensor of Crystals

Due to the lattice structure, crystals are electrically anisotropic. The constitutive relations for a homogeneous, lossless, and nonmagnetic crystal are[1]

$$\mathbf{D} = \varepsilon_0 \overline{\overline{\varepsilon}} \cdot \mathbf{E} \tag{3.1}$$

$$\mathbf{B} = \mu_0 \mathbf{H} \tag{3.2}$$

where $\overline{\overline{\varepsilon}}$ is a real constant dyadic or tensor of rank 2 and is called *dielectric tensor*. Equation (3.1) says that the electric flux density $\mathbf{D}$ is not in the direction of electric field intensity $\mathbf{E}$.
where

$$\overline{\overline{\varepsilon}} = \begin{bmatrix} \varepsilon_{11} & \varepsilon_{12} & \varepsilon_{13} \\ \varepsilon_{21} & \varepsilon_{22} & \varepsilon_{23} \\ \varepsilon_{31} & \varepsilon_{32} & \varepsilon_{33} \end{bmatrix} \tag{3.3}$$

The dielectric tensor $\overline{\overline{\varepsilon}}$ must be symmetric for a lossless crystal obeying with that the law of conservation of energy.

[1]where $\overline{\overline{\varepsilon}} = \varepsilon_r \overline{\overline{\mathbf{I}}}$.

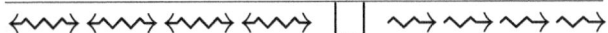

After reducing the real symmetric matrix to a diagonal form in an orthogonal coordinate system formed by eigenvectors, we get

$$\overline{\varepsilon} = \begin{bmatrix} \varepsilon_1 & 0 & 0 \\ 0 & \varepsilon_2 & 0 \\ 0 & 0 & \varepsilon_3 \end{bmatrix} \tag{3.4}$$

where the real constants ε_1, ε_2 and ε_3 are eigenvalues of ε and are called *principal dielectric constants.*
The determinant[2] of ε must always be positive and different from zero as a result of the positive definiteness of ε:

$$|\overline{\varepsilon}| = \varepsilon_1\varepsilon_2\varepsilon_3 > 0 \tag{3.5}$$

Hence, ε is nonsingular and thus the inverse of it always exists, so that (3.1) may be also written as

$$\begin{aligned} \overline{\varepsilon}^{\,-1} \cdot \mathbf{D} &= \varepsilon_0 \mathbf{E} \\ \frac{\overline{\varepsilon}^{\,-1} \cdot \mathbf{D}}{\varepsilon_0} &= \mathbf{E} \end{aligned} \tag{3.6}$$

3.1.2. Dispersion Equation

For nonmagnetic lossless crystals we have

$$\omega \mathbf{D}_0 = \omega\varepsilon_0\overline{\varepsilon} \cdot \mathbf{E}_0 = -\mathbf{k} \times \mathbf{H}_0 \tag{3.7}$$

$$\omega \mathbf{B}_0 = \omega\mu_0 \mathbf{H}_0 = \mathbf{k} \times \mathbf{E}_0 \tag{3.8}$$

With[3] the aid of antisymmetric matrix $\mathbf{k} \times \overline{\mathbf{I}}$

$$(\mathbf{k} \times \overline{\mathbf{I}}) = (\overline{\mathbf{I}} \times \mathbf{k}) = \begin{bmatrix} 0 & -k_3 & k_2 \\ k_3 & 0 & -k_1 \\ -k_2 & k_1 & 0 \end{bmatrix} \tag{3.9}$$

We have

$$\omega \mathbf{D}_0 = \omega\varepsilon_0\overline{\varepsilon} \cdot \mathbf{E}_0 = -(\mathbf{k} \times \overline{\mathbf{I}}) \cdot \mathbf{H}_0 \tag{3.10}$$

$$\omega \mathbf{B}_0 = \omega\mu_0 \mathbf{H}_0 = (\mathbf{k} \times \overline{\mathbf{I}}) \cdot \mathbf{E}_0 \tag{3.11}$$

Equations (3.10) and (3.11) are mutually coupled simultaneous equations. To decouple them, we eliminate $\mathbf{H}_0$ and obtain an equation in $\mathbf{E}_0$ alone:

$$\omega\varepsilon_0\overline{\varepsilon} \cdot \mathbf{E}_0 = -(\mathbf{k} \times \overline{\mathbf{I}}) \cdot \frac{(\mathbf{k} \times \overline{\mathbf{I}}) \cdot \mathbf{E}_0}{\omega\mu_0}$$

[2] See Appendix G2
[3] where $\mathbf{u} \times \mathbf{v} = (\mathbf{u} \times \overline{\mathbf{I}}) \cdot \mathbf{v} = (\overline{\mathbf{I}} \times \mathbf{u}) \cdot \mathbf{v}$

$$[k_0^2\overline{\varepsilon} + (\mathbf{k} \times \overline{\mathbf{I}})^2] \cdot \mathbf{E}_0 = \mathbf{0} \tag{3.12}$$

Dot-premultiplying both sides by ε^{-1}, we get

$$[k_0^2\overline{\mathbf{I}} + \overline{\varepsilon}^{\,-1} \cdot (\mathbf{k} \times \overline{\mathbf{I}})^2] \cdot \mathbf{E}_0 = \mathbf{0} \tag{3.13}$$

Equation (3.13) can be expressed in another form taking into consideration equation (3.6)

$$[k_0^2\overline{\mathbf{I}} + (\mathbf{k} \times \overline{\mathbf{I}})^2 \cdot \overline{\varepsilon}^{\,-1}] \cdot \mathbf{D}_0 = \mathbf{0} \tag{3.14}$$

Finally, eliminating $\mathbf{E}_0$ from equations (3.10) and (3.11) we find

$$[k_0^2\overline{\mathbf{I}} + (\mathbf{k} \times \overline{\mathbf{I}}) \cdot \overline{\varepsilon}^{\,-1} \cdot (\mathbf{k} \times \overline{\mathbf{I}})] \cdot \mathbf{H}_0 = \mathbf{0} \tag{3.15}$$

For a uniform plane wave with $\mathbf{k} = k\widehat{\mathbf{k}}$, we may write equation (3.13) in the form of an eigenvalue problem,

$$[\overline{\varepsilon}^{\,-1} \cdot (\widehat{\mathbf{k}} \times \overline{\mathbf{I}})^2] \cdot \mathbf{E}_0 = \lambda \mathbf{E}_0 \tag{3.16}$$

where $\lambda = \frac{-k_0^2}{k^2}$ is an eigenvalue of the $\overline{\varepsilon}^{\,-1} \cdot (\mathbf{k} \times \overline{\mathbf{I}})^2$ and $\mathbf{E}_0$ matrix is the corresponding eigenvector. Similarly, equations (3.14) and (3.15) may be written as

$$[(\widehat{\mathbf{k}} \times \overline{\mathbf{I}})^2 \cdot \overline{\varepsilon}^{\,-1}] \cdot \mathbf{D}_0 = \lambda \mathbf{D}_0 \tag{3.17}$$

and

$$[(\widehat{\mathbf{k}} \times \overline{\mathbf{I}}) \cdot \overline{\varepsilon}^{\,-1} \cdot (\widehat{\mathbf{k}} \times \overline{\mathbf{I}})] \cdot \mathbf{H}_0 = \lambda \mathbf{H}_0 \tag{3.18}$$

respectively.

For the homogeneous equation (3.13) to have a nonzero vector solution $\mathbf{E}_0$, it is necessary that the determinant of the coefficient matrix vanishes, i.e.,

$$|k_0^2\overline{\mathbf{I}} + \overline{\varepsilon}^{\,-1} \cdot (\mathbf{k} \times \overline{\mathbf{I}})^2| = 0 \tag{3.19}$$

This is the dispersion equation. Similarly, for nonzero $\mathbf{D}_0$ and $\mathbf{H}_0$ to exist, we must have

$$|k_0^2\overline{\mathbf{I}} + (\mathbf{k} \times \overline{\mathbf{I}})^2 \cdot \overline{\varepsilon}^{\,-1}| = 0 \tag{3.20}$$

$$|k_0^2\overline{\mathbf{I}} + (\mathbf{k} \times \overline{\mathbf{I}}) \cdot \overline{\varepsilon}^{\,-1} \cdot (\mathbf{k} \times \overline{\mathbf{I}})| = 0 \tag{3.21}$$

and

$$\begin{aligned}|k_0^2\overline{\mathbf{I}} + \varepsilon^{-1} \cdot (\mathbf{k} \times \overline{\mathbf{I}})^2| & \\ &= |k_0^2\overline{\mathbf{I}} + (\mathbf{k} \times \overline{\mathbf{I}})^2 \cdot \overline{\varepsilon}^{\,-1}| \\ &= |k_0^2\overline{\mathbf{I}} + (\mathbf{k} \times \overline{\mathbf{I}}) \cdot \overline{\varepsilon}^{\,-1} \cdot (\mathbf{k} \times \overline{\mathbf{I}})| \\ &= 0 \end{aligned} \tag{3.22}$$

We can also obtain the identity

$$\begin{aligned}|k_0^2\overline{\mathbf{I}} + \overline{\varepsilon}^{\,-1} \cdot (\mathbf{k} \times \overline{\mathbf{I}})^2| &= k_0^6 + [\overline{\varepsilon}^{\,-1} \cdot (\mathbf{k} \times \overline{\mathbf{I}})^2]_t k_0^4 \\ &+ \{\mathrm{adj}[\overline{\varepsilon}^{\,-1} \cdot (\mathbf{k} \times \overline{\mathbf{I}})^2]\}_t k_0^2 \\ &+ |\overline{\varepsilon}^{\,-1} \cdot (\mathbf{k} \times \overline{\mathbf{I}})^2| \end{aligned} \tag{3.23}$$

with the help of[4]

$$\mathrm{adj}(\overline{\mathbf{A}} \cdot \overline{\mathbf{B}}) = \mathrm{adj}\overline{\mathbf{B}} \cdot \mathrm{adj}\overline{\mathbf{A}} \tag{3.24}$$

$$(\mathbf{u} \times \overline{\mathbf{I}})^2 = \mathbf{u}\mathbf{u} - \mathbf{u}^2\overline{\mathbf{I}} \tag{3.25}$$

$$\mathrm{adj}(\mathbf{u} \times \overline{\mathbf{I}}) = \mathbf{u}\mathbf{u} \tag{3.26}$$

$$\mathrm{adj}(\overline{\mathbf{A}})^{-1} = (\mathrm{adj}\overline{\mathbf{A}})^{-1} = \frac{\overline{\mathbf{A}}}{|\overline{\mathbf{A}}|} \tag{3.27}$$

$$|\mathbf{u} \times \overline{\mathbf{I}}| = 0 \tag{3.28}$$

$$(\mathbf{u}\mathbf{v})_t = \mathbf{u} \cdot \mathbf{v} \tag{3.29}$$

And if

$$\overline{\mathbf{C}} = \overline{\mathbf{A}} \pm \lambda\overline{\mathbf{I}}$$

then [5]

$$\begin{aligned}
&(a) \quad \overline{\mathbf{C}}_t = \overline{\mathbf{A}}_t \pm 3\lambda \\
&(b) \quad |\overline{\mathbf{C}}| = \pm\lambda^3 + \overline{\mathbf{A}}_t\lambda^2 \pm (\mathrm{adj}\overline{\mathbf{A}})_t\lambda + |\overline{\mathbf{A}}| \\
&(c) \quad \mathrm{adj}\overline{\mathbf{C}} = \lambda^2\overline{\mathbf{I}} \pm \lambda(\overline{\mathbf{A}}_t\overline{\mathbf{I}} - \overline{\mathbf{A}}) + \mathrm{adj}\overline{\mathbf{A}} \\
&(d) \quad (\mathrm{adj}\overline{\mathbf{C}})_t = 3\lambda^2 \pm 2\overline{\mathbf{A}}_t\lambda + (\mathrm{adj}\overline{\mathbf{A}})_t
\end{aligned} \tag{3.30}$$

[4] The notation for the matrix of rank 2 tensor is to be distinguished using the small dash over it from the normal vector which is written without the dash.

[5] t corresponds with the trace of the matrix which is defined as $\overline{\mathbf{A}}_t = a_{ii} = a_{11} + a_{22} + a_{33}$

Solving for each term in equation (3.23), we get from equations (3.25) and (3.29)

$$[\overline{\varepsilon}^{\,-1}\cdot(\mathbf{k}\times\overline{\mathbf{I}})^2]_t \;=\; $$

$$= \;(\overline{\varepsilon}^{\,-1}\cdot\mathbf{kk} - k^2\overline{\varepsilon}^{\,-1})_t$$

using (3.27)

$$= \;\frac{1}{|\overline{\varepsilon}|}\mathbf{k}\cdot[\mathrm{adj}\overline{\varepsilon} - (\mathrm{adj}\overline{\varepsilon})_t\overline{\mathbf{I}}]\cdot\mathbf{k} \qquad (3.31)$$

The second term in equation (3.23) with the help of equations (3.24), (3.26) and (3.29) can be expressed as

$$\{\mathrm{adj}[\overline{\varepsilon}^{\,-1}\cdot(\mathbf{k}\times\overline{\mathbf{I}})^2]\}_t \;=\; [\mathrm{adj}(\mathbf{k}\times\overline{\mathbf{I}})\cdot\mathrm{adj}(\mathbf{k}\times\overline{\mathbf{I}})\cdot\mathrm{adj}\overline{\varepsilon}^{\,-1}]_t$$

$$= \;\frac{k^2}{|\overline{\varepsilon}|}(\mathbf{k}\cdot\overline{\varepsilon}\cdot\mathbf{k}) \qquad (3.32)$$

And the third term in equation (3.23) becomes

$$|\overline{\varepsilon}^{\,-1}\cdot(\mathbf{k}\times\overline{\mathbf{I}})^2| = |\overline{\varepsilon}^{\,-1}|\;\underbrace{|(\mathbf{k}\times\overline{\mathbf{I}})|}_{=0}\;\underbrace{|(\mathbf{k}\times\overline{\mathbf{I}})|}_{=0} = 0 \qquad (3.33)$$

Substituting equations (3.31), (3.32) and (3.33) into equation (3.23), we obtain the final explicit form of the dispersion equation from equation (3.19), namely

$$\frac{k^2}{|\overline{\varepsilon}|}(\mathbf{k}\cdot\overline{\varepsilon}\cdot\mathbf{k})k_0^2 + \frac{k_0^4}{|\overline{\varepsilon}|}\mathbf{k}\cdot[\mathrm{adj}\overline{\varepsilon} - (\mathrm{adj}\overline{\varepsilon})_t\overline{\mathbf{I}}]\cdot\mathbf{k} + k_0^6 = 0$$

$$k^2(\mathbf{k}\cdot\overline{\varepsilon}\cdot\mathbf{k}) + k_0^2\mathbf{k}\cdot[\mathrm{adj}\overline{\varepsilon} - (\mathrm{adj}\overline{\varepsilon})_t\overline{\mathbf{I}}]\cdot\mathbf{k} + |\overline{\varepsilon}|k_0^4 = 0 \qquad (3.34)$$

or from $\mathrm{adj}\overline{\mathbf{A}} = \overline{\mathbf{A}}^2 - \overline{\mathbf{A}}_t\overline{\mathbf{A}} + (\mathrm{adj}\overline{\mathbf{A}})_t\overline{\mathbf{I}}$, we get

$$k^2(\mathbf{k}\cdot\overline{\varepsilon}\cdot\mathbf{k})$$

$$+ \;k_0^2\mathbf{k}\cdot[\overline{\varepsilon}^2 - \overline{\varepsilon}_t\overline{\varepsilon} + \cancel{(\mathrm{adj}\overline{\varepsilon})_t\overline{\mathbf{I}}} - \cancel{(\mathrm{adj}\overline{\varepsilon})_t\overline{\mathbf{I}}}]\cdot\mathbf{k}$$

$$+ \;|\overline{\varepsilon}|k_0^4 = 0$$

$$k^2(\mathbf{k}\cdot\overline{\varepsilon}\cdot\mathbf{k})$$

$$- \;k_0^2\mathbf{k}\cdot\overline{\varepsilon}\cdot(\overline{\varepsilon}_t\overline{\mathbf{I}} - \overline{\varepsilon})\cdot\mathbf{k} + |\overline{\varepsilon}|k_0^4 = 0 \qquad (3.35)$$

or in terms of the refractive index vector $\mathbf{n} = n\widehat{\mathbf{k}} = \mathbf{k}/k_0$

$$n^4(\widehat{\mathbf{k}}\cdot\overline{\varepsilon}\cdot\widehat{\mathbf{k}}) + n^2\widehat{\mathbf{k}}\cdot[\mathrm{adj}\overline{\varepsilon} - (\mathrm{adj}\overline{\varepsilon})_t\overline{\mathbf{I}}]\cdot\widehat{\mathbf{k}} + |\overline{\varepsilon}| = 0 \qquad (3.36)$$

where $n = k/k_0$.

In an anisotropic medium, the index of refraction depends on the direction of wave normal, and the dispersion equation alone does not uniquely determine the vector **n**; this will be discussed later in this chapter.

For a given direction of wave normal $\widehat{\mathbf{k}}$, the two values of n^2 obtained from the dispersion equation (3.36) are always real and positive, and to prove this we dot-premultiply equation (3.13) by $\mathbf{E}_0^*$ and obtain

$$n^2 = \frac{\mathbf{E}_0^* \cdot \overline{\varepsilon} \cdot \mathbf{E}_0}{(\widehat{\mathbf{k}} \times \mathbf{E}_0)^* \cdot (\widehat{\mathbf{k}} \times \mathbf{E}_0)}$$

Since $\overline{\varepsilon}$ is a positive definite symmetric matrix, the numerator is always real and positive and hence $n^2 > 0$ for any complex $\mathbf{E}_0$.

There are some other remarks on the equations which can be taken into observation as well, such as looking on the homogeneous equation (3.12), which satisfies the condition for the existence of nonzero $\mathbf{E}_0$ in the dispersion equation

$$|k_0^2\overline{\varepsilon} + (\mathbf{k} \times \overline{\mathbf{I}})^2| = 0$$

But first of all, let's investigate the following identities:
If

$$\overline{\mathbf{C}} = \overline{\mathbf{A}} + \mathbf{uv}$$

then

$$(a) \quad \overline{\mathbf{C}}_t = \overline{\mathbf{A}}_t + \mathbf{u} \cdot \mathbf{v}$$

$$(b) \quad |\overline{\mathbf{C}}| = |\overline{\mathbf{A}}| + \mathbf{v} \cdot (\mathrm{adj}\overline{\mathbf{A}}) \cdot \mathbf{u}$$

$$(c) \quad \mathrm{adj}\overline{\mathbf{C}} = \mathrm{adj}\overline{\mathbf{A}} + (\overline{\mathbf{A}} - \overline{\mathbf{A}}_t\overline{\mathbf{I}}) \cdot (\mathbf{v} \times \overline{\mathbf{I}}) \cdot (\mathbf{u} \times \overline{\mathbf{I}}) + [(\mathbf{v} \cdot \overline{\mathbf{A}}) \times \overline{\mathbf{I}}] \cdot (\mathbf{u} \times \overline{\mathbf{I}}) \qquad (3.37)$$

$$(d) \quad (\mathrm{adj}\overline{\mathbf{C}})_t = (\mathrm{adj}\overline{\mathbf{A}})_t + (\mathbf{u} \cdot \mathbf{v})\overline{\mathbf{A}}_t - \mathbf{v} \cdot \overline{\mathbf{A}} \cdot \mathbf{u}$$

Using equation (3.25) and (3.37b), we obtain for the above mentioned dispersion equation the following

$$
\begin{aligned}
& |k_0^2\overline{\varepsilon} + (\mathbf{k} \times \overline{\mathbf{I}})^2| \\
= \quad & |(k_0^2\overline{\varepsilon} - k^2\overline{\mathbf{I}}) + \mathbf{k}\mathbf{k}| \\
= \quad & |(k_0^2\overline{\varepsilon} - k^2\overline{\mathbf{I}})| + \mathbf{k} \cdot [\mathrm{adj}(k_0^2\overline{\varepsilon} - k^2\overline{\mathbf{I}})] \cdot \mathbf{k} \\
= \quad & 0 \qquad (3.38)
\end{aligned}
$$

But

$$
\begin{aligned}
& |(k_0^2\overline{\varepsilon} - k^2\overline{\mathbf{I}})| \\
= \quad & (\widehat{\mathbf{k}} \cdot \overline{\mathbf{I}} \cdot \widehat{\mathbf{k}}) \ |(k_0^2\overline{\varepsilon} - k^2\overline{\mathbf{I}})| \\
= \quad & \widehat{\mathbf{k}} \cdot (|(k_0^2\overline{\varepsilon} - k^2\overline{\mathbf{I}})|\overline{\mathbf{I}}) \cdot \widehat{\mathbf{k}} \\
= \quad & \widehat{\mathbf{k}} \cdot [(k_0^2\overline{\varepsilon} - k^2\overline{\mathbf{I}}) \cdot \mathrm{adj}(k_0^2\overline{\varepsilon} - k^2\overline{\mathbf{I}})] \cdot \widehat{\mathbf{k}} \qquad (3.39)
\end{aligned}
$$

Substituting in (3.38)

$$
\begin{aligned}
& |k_0^2\overline{\varepsilon} + (\mathbf{k} \times \overline{\mathbf{I}})^2| \\
= \quad & \widehat{\mathbf{k}} \cdot [(k_0^2\overline{\varepsilon} - k^2\overline{\mathbf{I}}) \cdot \mathrm{adj}(k_0^2\overline{\varepsilon} - k^2\overline{\mathbf{I}})] \cdot \widehat{\mathbf{k}} \\
& + \mathbf{k} \cdot [\mathrm{adj}(k_0^2\overline{\varepsilon} - k^2\overline{\mathbf{I}})] \cdot \mathbf{k} \\
= \quad & - \widehat{\mathbf{k}} \cdot k^2\overline{\mathbf{I}} \cdot [\mathrm{adj}(k_0^2\overline{\varepsilon} - k^2\overline{\mathbf{I}})] \cdot \widehat{\mathbf{k}} + \mathbf{k} \cdot [\mathrm{adj}(k_0^2\overline{\varepsilon} - k^2\overline{\mathbf{I}})] \cdot \mathbf{k} \\
& + \widehat{\mathbf{k}} \cdot [(k_0^2\overline{\varepsilon}) \cdot \mathrm{adj}(k_0^2\overline{\varepsilon} - k^2\overline{\mathbf{I}})] \cdot \widehat{\mathbf{k}} \\
= \quad & \widehat{\mathbf{k}} \cdot \overline{\varepsilon} \cdot [\mathrm{adj}(k_0^2\overline{\varepsilon} - k^2\overline{\mathbf{I}})] \cdot \widehat{\mathbf{k}} \\
= \quad & 0 \qquad (3.40)
\end{aligned}
$$

If $k_0^2\overline{\varepsilon} - k^2\overline{\mathbf{I}}$ is nonsingular (that is, $|k_0^2\overline{\varepsilon} - k^2\overline{\mathbf{I}}| \neq 0$), we may divide equation (3.38) by $|k_0^2\overline{\varepsilon} - k^2\overline{\mathbf{I}}|$ and obtain the dispersion equation in the form of

$$1 + \mathbf{k} \cdot (k_0^2\overline{\varepsilon} - k^2\overline{\mathbf{I}})^{-1} \cdot \mathbf{k} = 0 \qquad (3.41)$$

Substitution of $\overline{\varepsilon} = |\overline{\varepsilon}| \ \mathrm{adj} \ \overline{\varepsilon}^{\ -1}$ into equation (3.40) yields another form:

$$\widehat{\mathbf{k}} \cdot \overline{\varepsilon} \cdot [\mathrm{adj}(k_0^2\overline{\varepsilon} - k^2\overline{\mathbf{I}})] \cdot \widehat{\mathbf{k}} = 0$$

$$\widehat{\mathbf{k}} \cdot |\overline{\varepsilon}| \ \mathrm{adj} \ \overline{\varepsilon}^{\ -1} \cdot [\mathrm{adj}(k_0^2\overline{\varepsilon} - k^2\overline{\mathbf{I}})] \cdot \widehat{\mathbf{k}} = 0$$

And using equation (3.24)

$$\widehat{\mathbf{k}} \cdot \mathrm{adj}\bar{\varepsilon}^{\,-1} \cdot [\mathrm{adj}(k_0^2\bar{\varepsilon} - k^2\bar{\mathbf{I}})] \cdot \widehat{\mathbf{k}} = 0$$
$$\widehat{\mathbf{k}} \cdot [\mathrm{adj}(k_0^2\bar{\varepsilon} \cdot \bar{\varepsilon}^{\,-1} - k^2\bar{\mathbf{I}} \cdot \bar{\varepsilon}^{\,-1})] \cdot \widehat{\mathbf{k}} = 0$$
$$\widehat{\mathbf{k}} \cdot [\mathrm{adj}(\bar{\varepsilon}^{\,-1} - \frac{1}{n^2}\bar{\mathbf{I}})] \cdot \widehat{\mathbf{k}} = 0 \qquad (3.42)$$

If $|k_0^2\bar{\varepsilon} - k^2\bar{\mathbf{I}}| \neq 0$ thus $|\bar{\varepsilon}^{\,-1} - 1/n^2\bar{\mathbf{I}}| \neq 0$, we may divide equations (3.40) and (3.42) by these determinants respectively and obtain still two other forms of the dispersion equation

$$\widehat{\mathbf{k}} \cdot \bar{\varepsilon} \cdot (k_0^2\bar{\varepsilon} - k^2\bar{\mathbf{I}})^{-1} \cdot \widehat{\mathbf{k}} = 0 \qquad (3.43)$$
$$\widehat{\mathbf{k}} \cdot (\bar{\varepsilon}^{\,-1} - \frac{1}{n^2}\bar{\mathbf{I}})^{-1} \cdot \widehat{\mathbf{k}} = 0 \qquad (3.44)$$

With respect to the principal dielectric axes of the tensor $\bar{\varepsilon}$, we may write

$$\bar{\varepsilon}^{\,-1} - \frac{1}{n^2}\bar{\mathbf{I}} = \begin{bmatrix} \frac{1}{\varepsilon_1} - \frac{1}{n^2} & 0 & 0 \\ 0 & \frac{1}{\varepsilon_2} - \frac{1}{n^2} & 0 \\ 0 & 0 & \frac{1}{\varepsilon_3} - \frac{1}{n^2} \end{bmatrix} \qquad (3.45)$$

Let l_1, l_2 and l_3 be the direction cosines of the wave normal $\widehat{\mathbf{k}}$, that is,

$$\widehat{\mathbf{k}} = \begin{bmatrix} l_1 \\ l_2 \\ l_3 \end{bmatrix} \qquad (3.46)$$

with $l_1^2 + l_2^2 + l_3^2 = 1$. Substitution of equations (3.46) and (3.45) into equation (3.44) yields

$$\frac{l_1^2}{\frac{1}{\varepsilon_1} - \frac{1}{n^2}} + \frac{l_2^2}{\frac{1}{\varepsilon_2} - \frac{1}{n^2}} + \frac{l_3^2}{\frac{1}{\varepsilon_3} - \frac{1}{n^2}} = 0 \qquad (3.47)$$

Which is one of the generally used forms of dispersion equation.

3.2. Polarization of Waves in Crystals, Optic Axis

The complex conjugates of equations (3.7) and (3.8) are

$$\omega \mathbf{D}_0^* = \omega \varepsilon_0 \overline{\varepsilon} \cdot \mathbf{E}_0^* = -\mathbf{k} \times \mathbf{H}_0^* \tag{3.48}$$

$$\omega \mathbf{B}_0^* = \omega \mu_0 \mathbf{H}_0^* = \mathbf{k} \times \mathbf{E}_0^* \tag{3.49}$$

Noticing that

$$(\mathrm{adj}\overline{\mathbf{A}}) \cdot (\mathbf{u} \times \mathbf{v}) = (\mathbf{u} \cdot \overline{\mathbf{A}}) \times (\mathbf{v} \cdot \overline{\mathbf{A}}) \tag{3.50}$$

we cross-multiply equations (3.7) and (3.48) to obtain

$$\begin{aligned} \mathbf{D}_0 \times \mathbf{D}_0^* &= \varepsilon_0^2 (\overline{\varepsilon} \cdot \mathbf{E}_0) \times (\overline{\varepsilon} \cdot \mathbf{E}_0^*) \\ &= \varepsilon_0^2 (\mathrm{adj}\overline{\varepsilon}) \cdot (\mathbf{E}_0 \times \mathbf{E}_0^*) \end{aligned} \tag{3.51}$$

or [6]

$$\begin{aligned} \mathbf{D}_0 \times \mathbf{D}_0^* &= \frac{1}{\omega^2} (\mathbf{k} \times \mathbf{H}_0) \times (\mathbf{k} \times \mathbf{H}_0^*) \\ &= \frac{1}{\omega^2} \mathbf{k}\mathbf{k} \cdot (\mathbf{H}_0 \times \mathbf{H}_0^*) \end{aligned} \tag{3.52}$$

Hence,

$$\mathbf{E}_0 \times \mathbf{E}_0^* = \frac{1}{\omega^2 \varepsilon_0^2 |\overline{\varepsilon}|} \overline{\varepsilon} \cdot \mathbf{k}\mathbf{k} \cdot (\mathbf{H}_0 \times \mathbf{H}_0^*) \tag{3.53}$$

Similarly, cross-multiplying equations (3.8) and (3.49) we find

$$\begin{aligned} \mathbf{H}_0 \times \mathbf{H}_0^* &= \frac{1}{\omega^2 \mu_0^2} (\mathbf{k} \times \mathbf{E}_0) \times (\mathbf{k} \times \mathbf{E}_0^*) \\ &= \frac{1}{\omega^2 \mu_0^2} \mathbf{k}\mathbf{k} \cdot (\mathbf{E}_0 \times \mathbf{E}_0^*) \end{aligned} \tag{3.54}$$

[6] $(\mathbf{k} \times \mathbf{H}_0) \times (\mathbf{k} \times \mathbf{H}_0^*)$

$= (\mathbf{k} \times \mathbf{H}_0) \cdot \mathbf{H}_0^* \mathbf{k} - \underbrace{(\mathbf{k} \times \mathbf{H}_0) \cdot \mathbf{k}}_{=0} \mathbf{H}_0^* = \mathbf{k} \cdot (\mathbf{H}_0 \times \mathbf{H}_0^*) \mathbf{k}$

Eliminating either $(\mathbf{H}_0 \times \mathbf{H}_0^*)$ or $(\mathbf{E}_0 \times \mathbf{E}_0^*)$ from equations (3.53) and (3.54) we obtain

$$(\bar{\mathbf{I}} - \frac{\mathbf{k}^2}{k_0^4|\bar{\varepsilon}|}\bar{\varepsilon} \cdot \mathbf{kk}) \cdot (\mathbf{E}_0 \times \mathbf{E}_0^*) = \mathbf{0} \tag{3.55}$$

$$(\bar{\mathbf{I}} - \frac{\mathbf{k} \cdot \bar{\varepsilon} \cdot \mathbf{k}}{k_0^4|\bar{\varepsilon}|} \cdot \mathbf{kk}) \cdot (\mathbf{H}_0 \times \mathbf{H}_0^*) = \mathbf{0} \tag{3.56}$$

Equations (3.55) and (3.56) are two homogeneous equations. Thus for nonzero vector $(\mathbf{E}_0 \times \mathbf{E}_0^*)$ or $(\mathbf{H}_0 \times \mathbf{H}_0^*)$ to exist, the determinant of the coefficient matrix of equation (3.55) or equation (3.56) must vanish; i.e.

$$|\bar{\mathbf{I}} - \frac{\mathbf{k}^2}{k_0^4|\bar{\varepsilon}|}\bar{\varepsilon} \cdot \mathbf{kk}| = 0 \tag{3.57}$$

or

$$|\bar{\mathbf{I}} - \frac{\mathbf{k} \cdot \bar{\varepsilon} \cdot \mathbf{k}}{k_0^4|\bar{\varepsilon}|} \cdot \mathbf{kk}| = 0 \tag{3.58}$$

If $\overline{\mathbf{C}} = \lambda\bar{\mathbf{I}} + \mathbf{uv}$,

$$\begin{aligned}
&(a) \quad \overline{\mathbf{C}}_t = 3\lambda + \mathbf{u} \cdot \mathbf{v} \\
&(b) \quad |\overline{\mathbf{C}}| = \lambda^2(\lambda + \mathbf{u} \cdot \mathbf{v}) \\
&(c) \quad \mathrm{adj}\overline{\mathbf{C}} = \lambda[(\lambda + \mathbf{u} \cdot \mathbf{v})\bar{\mathbf{I}} - \mathbf{uv}] \\
&(d) \quad \mathrm{adj}(\overline{\mathbf{C}})_t = \lambda(3\lambda + 2\mathbf{u} \cdot \mathbf{v})
\end{aligned} \tag{3.59}$$

Calculating the determinants in equations (3.57) and (3.58) according to equation (3.59b) we find the both conditions give the same results

$$(\mathbf{k} \cdot \bar{\varepsilon} \cdot \mathbf{k})\mathbf{k}^2 - k_0^4|\bar{\varepsilon}| = 0 \tag{3.60}$$

$$(\widehat{\mathbf{k}} \cdot \bar{\varepsilon} \cdot \widehat{\mathbf{k}})n^4 - |\bar{\varepsilon}| = 0 \tag{3.61}$$

which means, $(\mathbf{E}_0 \times \mathbf{E}_0^*)$, $(\mathbf{H}_0 \times \mathbf{H}_0^*)$ and $(\mathbf{D}_0 \times \mathbf{D}_0^*)$ are either all zero or nonzero depending on whether or not the wave vector $\mathbf{k}$ satisfies equation (3.60).

The uniform plane waves in a lossless crystal are always linearly polarized [7] except when the wave vector $\mathbf{k}$ satisfies the condition (3.60). In

[7] when $(\mathbf{E}_0 \times \mathbf{E}_0^*)=0$ then we have a linearly polarized wave.

the latter case, the wave can have any polarization. But the wave vector $\mathbf{k} = k_0\mathbf{n}$ must also satisfy the dispersion equation (3.36), the solutions of which are

$$n^2 = \frac{\widehat{\mathbf{k}} \cdot [(\text{adj}\overline{\varepsilon})_t\overline{\mathbf{I}} - \text{adj}\overline{\varepsilon}] \cdot \widehat{\mathbf{k}} \pm \sqrt{\Delta}}{2(\widehat{\mathbf{k}} \cdot \overline{\varepsilon} \cdot \widehat{\mathbf{k}})} \tag{3.62}$$

where the discriminant Δ is given by

$$\Delta = \{\widehat{\mathbf{k}} \cdot [(\text{adj}\overline{\varepsilon})_t\overline{\mathbf{I}} - \text{adj}\overline{\varepsilon}] \cdot \widehat{\mathbf{k}}\}^2 - 4(\widehat{\mathbf{k}} \cdot \overline{\varepsilon} \cdot \widehat{\mathbf{k}})|\overline{\varepsilon}| \tag{3.63}$$

Condition (3.61) and the dispersion equation (3.36) imply that

$$\widehat{\mathbf{k}} \cdot [(\text{adj}\overline{\varepsilon})_t\overline{\mathbf{I}} - \text{adj}\overline{\varepsilon}] \cdot \widehat{\mathbf{k}}n^2 = 2|\overline{\varepsilon}| \tag{3.64}$$

and substitution of equations (3.61) and (3.64) into equation (3.63) yields

$$\Delta = 0 \tag{3.65}$$

Which means that, if the refractive index $\mathbf{n}$ satisfies the condition (3.61), the two solutions in n^2 of the dispersion equation become equal. And the directions of the wave normal that cause the discriminant (3.65) to vanish and yield two equal roots for n^2 are called *optic axes* of the crystals.

The *optic axes* in crystals are the directions where the two values of the phase velocity $v_p = c/n = c/\sqrt{\varepsilon_r\mu_r}$ are equal[8]. If a crystal possesses two optic axes, it is said to be *biaxial*; if it possesses only one optic axis, it is *uniaxial*.

Aside from the optic axis, for each given direction of wave normal $\widehat{\mathbf{k}}$ there are two linearly polarized waves propagating at different phase velocities in the crystals as equation (3.55) says, either the term $(\overline{\mathbf{I}} - \frac{\mathbf{k}^2}{k_0^4|\overline{\varepsilon}|}\overline{\varepsilon} \cdot \mathbf{kk})$ equals to zero or $(\mathbf{E}_0 \times \mathbf{E}_0^*)$, and when the latter is the one which equals to zero, then it means that definitely the wave is linearly polarized, if it's not equal to zero then it means that the wave is definitely not linearly polarized.

In the case when the waves are linearly polarized, we may assume that all amplitude vectors are real. Hence, according to Maxwell's equations the vector $\mathbf{H}_0$ (and hence also $\mathbf{B}_0$) is perpendicular to $\mathbf{E}_0$, $\mathbf{D}_0$ and $\widehat{\mathbf{k}}$, which must therefore be coplanar[9]. From equations (3.7) and (3.8) we have

$$\widehat{\mathbf{k}} \cdot \mathbf{D}_0 = \varepsilon_0\widehat{\mathbf{k}} \cdot \overline{\varepsilon} \cdot \mathbf{E}_0 = \varepsilon_0\mathbf{E}_0 \cdot \overline{\varepsilon} \cdot \widehat{\mathbf{k}} = 0 \tag{3.66}$$

that is, unlike isotropic media, the electric field intensity is not perpendicular to $\widehat{\mathbf{k}}$, but to the vector $(\overline{\varepsilon} \cdot \widehat{\mathbf{k}})$; see Figure 3.1 below.

[8] where c: is the speed of light in free space.
[9] See Appendix G1

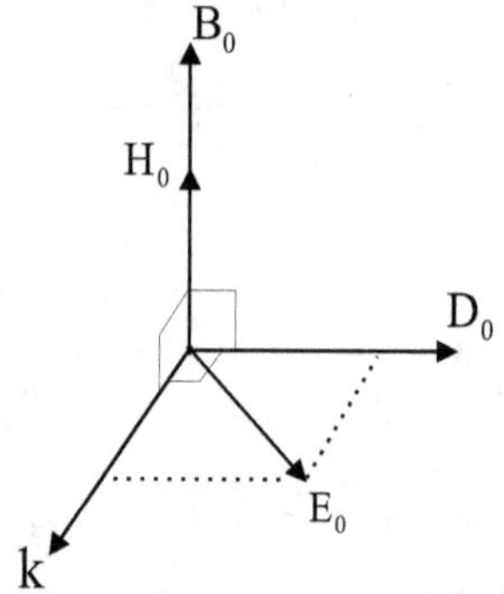

Figure 3.1.: Orientations of the Field Vectors and the Wave Vector.

3.3. Determination of Wave and Field Vectors

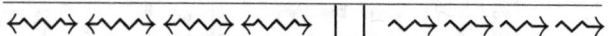

When any of the vectors $\mathbf{E}_0$, $\mathbf{D}_0$, $\mathbf{H}_0$ or $(\mathbf{B}_0 = \mu_0 \mathbf{H}_0)$, and $\mathbf{k}$ is given, the remaining vectors of the uniform plane wave in crystals are completely determined. When either $\mathbf{E}_0$ or $\mathbf{D}_0 = \varepsilon_0 \bar{\bar{\varepsilon}} \cdot \mathbf{E}_0$ is given, we take the dot product of (3.7) with $\mathbf{E}_0$ and using equation (3.8) we find

$$\mathbf{E}_0 \cdot \mathbf{D}_0 = \frac{-1}{\omega} \mathbf{E}_0 \cdot (\mathbf{k} \times \mathbf{H}_0) \tag{3.67}$$

$$= \frac{1}{\omega} \mathbf{H}_0 \cdot (\mathbf{k} \times \mathbf{E}_0) \tag{3.68}$$

$$= \mu_0 \mathbf{H}_0^2 \tag{3.69}$$

or

$$\mathbf{H}_0^2 = \frac{1}{\mu_0} (\mathbf{E}_0 \cdot \mathbf{D}_0) \tag{3.70}$$

Since $\mathbf{E}_0$ is perpendicular to $\mathbf{H}_0$, and $\mathbf{H}_0$ is perpendicular to $\mathbf{D}_0$, thus $\mathbf{H}_0$ is parallel to $\mathbf{E}_0 \times \mathbf{D}_0$

$$\mathbf{H}_0 \parallel \mathbf{E}_0 \times \mathbf{D}_0 \tag{3.71}$$

On the other hand, taking the dot product of equation (3.7) with itself and using equation (3.70), we obtain

$$\mathbf{k}^2 = \frac{\omega^2 \mathbf{D}_0^2}{\mathbf{H}_0^2} = \frac{\omega^2 \mu_0 \mathbf{D}_0^2}{\mathbf{E}_0 \cdot \mathbf{D}_0} \tag{3.72}$$

To find the direction of $\mathbf{k}$, we note that $\mathbf{H}_0 \perp \widehat{\mathbf{k}}$ and $\widehat{\mathbf{k}} \perp \mathbf{D}_0$ thus

$$\widehat{\mathbf{k}} \parallel \mathbf{D}_0 \times \mathbf{H}_0 \parallel \mathbf{D}_0 \times (\mathbf{E}_0 \times \mathbf{D}_0) \tag{3.73}$$

In summary, for a given $\mathbf{E}_0$ or $\mathbf{D}_0$, we may determine the magnitude of $\mathbf{H}_0$ from equation (3.70) and its direction from equation (3.71), the magnitude of $\mathbf{k}$ from equation (3.72), and its direction from equation (3.73).

If $\mathbf{H}_0$ is given. As a result from Maxwell's equations, we see that $\mathbf{E}_0 \cdot \mathbf{H}_0 = 0$ and

$$\mathbf{H}_0 \cdot \mathbf{D}_0 = \varepsilon_0 \mathbf{H}_0 \cdot \overline{\varepsilon} \cdot \mathbf{E}_0 = \varepsilon_0 \mathbf{E}_0 \cdot \overline{\varepsilon} \cdot \mathbf{H}_0 = 0 \tag{3.74}$$

Consequently vector $\mathbf{E}_0$ is perpendicular to both vectors $\mathbf{H}_0$ and $\overline{\varepsilon} \cdot \mathbf{H}_0$, and hence parallel to their cross product:

$$\mathbf{E}_0 = C_1[\mathbf{H}_0 \times (\overline{\varepsilon} \cdot \mathbf{H}_0)] \tag{3.75}$$

where C_1 is an arbitrary constant. Substitution of the above equation into (5.62) yields

$$\begin{aligned} &\mathbf{k}^2 \\ = &\frac{\omega^2 \mu_0 \mathbf{D}_0^2}{\mathbf{E}_0 \cdot \mathbf{D}_0} \\ = &\frac{k_0^2 (\overline{\varepsilon} \cdot \mathbf{E}_0) \cdot (\overline{\varepsilon} \cdot \mathbf{E}_0)}{\mathbf{E}_0 \cdot \overline{\varepsilon} \cdot \mathbf{E}_0} \\ = &\frac{k_0^2 \{\overline{\varepsilon} \cdot [\mathbf{H}_0 \times (\overline{\varepsilon} \cdot \mathbf{H}_0)]\}^2}{[\mathbf{H}_0 \times (\overline{\varepsilon} \cdot \mathbf{H}_0)] \cdot \overline{\varepsilon} \cdot [\mathbf{H}_0 \times (\overline{\varepsilon} \cdot \mathbf{H}_0)]} \end{aligned} \tag{3.76}$$

Using the identity [10]

$$\overline{\mathbf{A}} \cdot (\mathbf{u} \times \overline{\mathbf{I}}) = \{[(\mathrm{adj}\widetilde{\overline{\mathbf{A}}}) \cdot \mathbf{u}] \times \overline{\mathbf{I}}\} \cdot \overline{\mathbf{A}}^{\,-1} \tag{3.77}$$

[10] $\widetilde{\overline{\mathbf{A}}}$ or $\overline{\mathbf{A}}^{\,T}$ is the transpose of the matrix.

we may write

$$\bar{\varepsilon} \cdot [\mathbf{H}_0 \times (\bar{\varepsilon} \cdot \mathbf{H}_0)] = [(\mathrm{adj}\bar{\varepsilon}) \cdot \mathbf{H}_0] \times \mathbf{H}_0 \tag{3.78}$$

Thus equation (3.76) may also be expressed as:

$$\mathbf{k}^2 = \quad = \frac{k_0^2\{[(\mathrm{adj}\bar{\varepsilon}) \cdot \mathbf{H}_0] \times \mathbf{H}_0\}^2}{[\mathbf{H}_0 \times (\bar{\varepsilon} \cdot \mathbf{H}_0)] \cdot [(\mathrm{adj}\bar{\varepsilon}) \cdot \mathbf{H}_0] \times \mathbf{H}_0} \tag{3.79}$$

Also

$$\widehat{\mathbf{k}} \parallel \mathbf{D}_0 \times \mathbf{H}_0 \parallel \{\mathbf{H}_0 \times [\mathbf{H}_0 \times (\mathrm{adj}\bar{\varepsilon}) \cdot \mathbf{H}_0]\} \tag{3.80}$$

Therefore, when $\mathbf{H}_0$ is given, we may find the magnitude of $\mathbf{k}$ from equation (3.76) or (3.79) and its direction from equation (3.80). The vector $\mathbf{D}_0$ can be determined from equation (3.7), and vector $\mathbf{E}_0$ can be determined from equation (3.6).

If the wave vector $\mathbf{k}$ is given, according to (3.12), we have

$$(k_0^2\bar{\varepsilon} - k^2\bar{\mathbf{I}}) \cdot \mathbf{E}_0 = -(\mathbf{k} \cdot \mathbf{E}_0)\mathbf{k} \tag{3.81}$$

or

$$(k_0^2\bar{\varepsilon} - k^2\bar{\mathbf{I}}) \cdot \mathbf{E}_0 \parallel \mathbf{k} \tag{3.82}$$

Thus the direction of $\mathbf{E}_0$ depends on whether $k_0^2\bar{\varepsilon} - k^2\bar{\mathbf{I}}$ is singular or nonsingular.

1. Matrix $k_0^2\bar{\varepsilon} - k^2\bar{\mathbf{I}}$ is *nonsingular*. In other words, we have

$$|k_0^2\bar{\varepsilon} - k^2\bar{\mathbf{I}}| \neq 0 \tag{3.83}$$

and the inverse exists. Dot premultiplication of equation (3.81) by the inverse of $k_0^2\bar{\varepsilon} - k^2\bar{\mathbf{I}}$ yields the direction $\mathbf{E}_0$:

$$\begin{aligned}
(k_0^2\bar{\varepsilon} - k^2\bar{\mathbf{I}})^{-1} \quad &\cdot \quad (k_0^2\bar{\varepsilon} - k^2\bar{\mathbf{I}}) \cdot \mathbf{E}_0 \\
&= -(k_0^2\bar{\varepsilon} - k^2\bar{\mathbf{I}})^{-1} \cdot (\mathbf{k} \cdot \mathbf{E}_0)\mathbf{k} \\
\bar{\mathbf{I}} \cdot \mathbf{E}_0 &= -\frac{|k_0^2\bar{\varepsilon} - k^2\bar{\mathbf{I}}|}{|k_0^2\bar{\varepsilon} - k^2\bar{\mathbf{I}}|}(k_0^2\bar{\varepsilon} - k^2\bar{\mathbf{I}})^{-1} \cdot (\mathbf{k} \cdot \mathbf{E}_0)\mathbf{k} \\
\frac{\mathbf{E}_0|k_0^2\bar{\varepsilon} - k^2\bar{\mathbf{I}}|}{(\mathbf{k} \cdot \mathbf{E}_0)} &= -|k_0^2\bar{\varepsilon} - k^2\bar{\mathbf{I}}|(k_0^2\bar{\varepsilon} - k^2\bar{\mathbf{I}})^{-1} \cdot \mathbf{k} \\
\mathbf{e} &\sim [\mathrm{adj}(k_0^2\bar{\varepsilon} - k^2\bar{\mathbf{I}})] \cdot \mathbf{k}
\end{aligned} \tag{3.84}$$

and the direction of $\mathbf{D}_0$ follows from the constitutive relation

$$\mathbf{d} \quad \sim \quad \varepsilon_0\overline{\varepsilon}\cdot\mathbf{e} \sim \varepsilon_0\overline{\varepsilon}\cdot[\mathrm{adj}(k_0^2\overline{\varepsilon} - k^2\overline{\mathbf{I}})]\cdot\mathbf{k} \tag{3.85}$$

If we take the dot product of equation (3.85) with $\mathbf{k}$ and note that $\mathbf{k}\cdot\mathbf{D}_0 = 0$, we obtain

$$\widehat{\mathbf{k}}\cdot\overline{\varepsilon}\cdot[\mathrm{adj}(k_0^2\overline{\varepsilon} - k^2\overline{\mathbf{I}})]\cdot\widehat{\mathbf{k}} = 0 \tag{3.86}$$

which is the dispersion equation (3.40).

The direction of $\mathbf{H}_0$ may be obtained from Maxwell equations

$$\mathbf{h} \sim \frac{1}{\omega\mu_0}(\mathbf{k}\times\mathbf{e}) = \frac{1}{\omega\mu_0}\mathbf{k}\times\{[\mathrm{adj}(k_0^2\overline{\varepsilon} - k^2\overline{\mathbf{I}})]\cdot\mathbf{k}\} \tag{3.87}$$

or alternatively

$$\mathbf{h} \sim \frac{\omega}{k^2}(\mathbf{k}\times\mathbf{d}) = \frac{\omega\varepsilon_0}{k^2}\mathbf{k}\times\{\overline{\varepsilon}\times[\mathrm{adj}(k_0^2\overline{\varepsilon} - k^2\overline{\mathbf{I}})]\cdot\mathbf{k}\} \tag{3.88}$$

As an alternative approach, we may begin with equation (3.15). Expanding the product $(\mathbf{k}\times\overline{\mathbf{I}})\cdot\overline{\varepsilon}^{\,-1}\cdot(\mathbf{k}\times\overline{\mathbf{I}})$ in equation (3.15) according to

$$(\mathbf{v}\times\overline{\mathbf{I}})\cdot\mathrm{adj}\overline{\mathbf{A}}\cdot(\mathbf{u}\times\overline{\mathbf{I}}) = \widetilde{\overline{\mathbf{A}}}\cdot\mathbf{uv}\cdot\widetilde{\overline{\mathbf{A}}} - (\mathbf{u}\cdot\overline{\mathbf{A}}\cdot\mathbf{v})\widetilde{\overline{\mathbf{A}}} \tag{3.89}$$

$$\frac{|\overline{\varepsilon}|}{|\overline{\varepsilon}|}(\mathbf{k}\times\overline{\mathbf{I}})\cdot\overline{\varepsilon}^{\,-1} \quad\cdot\quad (\mathbf{k}\times\overline{\mathbf{I}})$$

$$= \frac{(\overline{\varepsilon}^{\,-1}\cdot\mathbf{kk}\cdot\overline{\varepsilon}^{\,-1} - (\mathbf{k}\cdot\overline{\varepsilon}^{\,-1}\cdot\mathbf{k})\overline{\varepsilon}^{\,-1})|\overline{\varepsilon}|}{|\overline{\varepsilon}|}$$

$$\frac{1}{|\overline{\varepsilon}|}(\mathbf{k}\times\overline{\mathbf{I}}) \quad\cdot\quad \mathrm{adj}\overline{\varepsilon}\cdot(\mathbf{k}\times\overline{\mathbf{I}})$$

$$= \frac{\mathrm{adj}\overline{\varepsilon}\cdot\mathbf{kk}\cdot\overline{\varepsilon}^{\,-1} - (\mathbf{k}\cdot\overline{\varepsilon}^{\,-1}\cdot\mathbf{k})\mathrm{adj}\overline{\varepsilon}}{|\overline{\varepsilon}|}$$

and [11] then dot-premultiplying the resulted former equation by $\mathrm{adj}\overline{\varepsilon}$

$$= \frac{\mathbf{kk}\cdot\overline{\varepsilon}^{\,-1} - (\mathbf{k}\cdot\overline{\varepsilon}^{\,-1}\cdot\mathbf{k})\overline{\mathbf{I}}}{|\overline{\varepsilon}|}$$

[11] $\overline{\varepsilon}^{\,-1} = (\overline{\varepsilon}^{\,-1})^T$

then (3.15) will become

$$[k_0^2(\text{adj}\overline{\varepsilon}) - (\mathbf{k}\cdot\overline{\varepsilon}\cdot\mathbf{k})\overline{\mathbf{I}}]\cdot\mathbf{H}_0 = -(\mathbf{k}\cdot\overline{\varepsilon}\cdot\mathbf{H}_0)\mathbf{k} \tag{3.90}$$

Now if the matrix $[k_0^2(\text{adj}\overline{\varepsilon}) - (\mathbf{k}\cdot\overline{\varepsilon}\cdot\mathbf{k})\overline{\mathbf{I}}]$ is nonsingular, i.e., if

$$|k_0^2(\text{adj}\overline{\varepsilon}) - (\mathbf{k}\cdot\overline{\varepsilon}\cdot\mathbf{k})\overline{\mathbf{I}}| \neq 0 \tag{3.91}$$

the direction of $\mathbf{H}_0$ follows from equation (3.90)

$$\mathbf{h}' \sim \{\text{adj}[k_0^2(\text{adj}\overline{\varepsilon}) - (\mathbf{k}\cdot\overline{\varepsilon}\cdot\mathbf{k})\overline{\mathbf{I}}]\}\cdot\mathbf{k} \tag{3.92}$$

Thus

$$\begin{aligned}\mathbf{d}' &\sim \frac{-1}{\omega}(\mathbf{k}\times\mathbf{h}') \\ &= \frac{1}{\omega}\{\ [\text{adj}\ (k_0^2\text{adj}\overline{\varepsilon} - (\mathbf{k}\cdot\overline{\varepsilon}\cdot\mathbf{k})\overline{\mathbf{I}})\ ]\cdot\mathbf{k}\ \}\times\mathbf{k}\end{aligned} \tag{3.93}$$

The direction of $\mathbf{E}_0$ may be found from the constitutive relation

$$\begin{aligned}\mathbf{e}' &\sim \frac{\overline{\varepsilon}^{\ -1}\cdot\mathbf{d}'}{\varepsilon_0} \\ &= \frac{1}{\omega\varepsilon_0}\overline{\varepsilon}^{\ -1}\cdot\{\ [\text{adj}\ (k_0^2\text{adj}\overline{\varepsilon} - (\mathbf{k}\cdot\overline{\varepsilon}\cdot\mathbf{k})\overline{\mathbf{I}})\ ]\cdot\mathbf{k}\ \} \\ &\quad\times\mathbf{k}\end{aligned} \tag{3.94}$$

Furthermore, from the fact that $\mathbf{k}\cdot\mathbf{H}_0 = 0$ and equation (3.92) we obtain

$$\widehat{\mathbf{k}}\cdot\{\ \text{adj}\ (k_0^2\text{adj}\overline{\varepsilon} - (\mathbf{k}\cdot\overline{\varepsilon}\cdot\mathbf{k})\overline{\mathbf{I}})\ \}\cdot\widehat{\mathbf{k}} = 0 \tag{3.95}$$

This is again another form from the dispersion equation. We also note that dot-premultiplying equation (3.81) by $(k_0^2\overline{\varepsilon} - k^2\overline{\mathbf{I}})^{-1}$ would give us results which make last calculated equations also valid.

2. Matrix $k_0^2\overline{\varepsilon} - k^2\overline{\mathbf{I}}$ is *singular*. In other words, we have

$$|k_0^2\overline{\varepsilon} - k^2\overline{\mathbf{I}}| = 0 \tag{3.96}$$

which means that k^2 is an eigenvalue of $k_0^2\overline{\varepsilon}$. Because the inverse of the matrix $k_0^2\overline{\varepsilon} - k^2\overline{\mathbf{I}}$ does not exist, the method used to determine the direction of $\mathbf{E}_0$ fails.

Let $\mathbf{l}_i$ be the ith eigenvector of the matrix $k_0^2\bar{\varepsilon}$ with the corresponding eigenvalue k_i^2, that is,

$$k_0^2\ \bar{\varepsilon} \cdot \mathbf{l}_i = k_i^2 \mathbf{l}_i \qquad (no\ sum) \qquad (3.97)$$

where k_i^2 is a solution of equation (3.96). Since $(k_0^2\bar{\varepsilon} - k^2\bar{\mathbf{I}})$ is a symmetric matrix, according to [Chen (1983)], we may write

$$\text{adj}(k_0^2\ \bar{\varepsilon} - k_i^2 \mathbf{l}_i) = C\mathbf{l}_i\mathbf{l}_i \qquad (no\ sum) \qquad (3.98)$$

where C is an arbitrary constant. Substitution of equations (3.96) and (3.98) into the dispersion equation (3.38) yields the condition to be imposed on the direction of the wave vector $\mathbf{k}_i$:

$$\mathbf{k}_i \cdot \mathbf{l}_i = 0 \qquad (no\ sum) \qquad (3.99)$$

i.e., the wave vector $\mathbf{k}_i$ must lie on a plane perpendicular to the ith eigenvector of $k_0^2\ \bar{\varepsilon}$.

To determine the direction of $\mathbf{H}_0$, we dot-premultiply equation (3.90) by $\mathbf{l}_i$ and then use the condition (3.99) to obtain

$$k_0^2\mathbf{l}_i \cdot (\text{adj}\bar{\varepsilon}) \cdot \mathbf{H}_0 - (\mathbf{k}_i \cdot \bar{\varepsilon} \cdot \mathbf{k}_i)(\mathbf{l}_i \cdot \mathbf{H}_0) = 0 \qquad (3.100)$$

Dot premultiplication of equation (3.97) by $\bar{\varepsilon}^{\ -1}$ gives

$$\mathbf{l}_i \cdot (\text{adj}\bar{\varepsilon}) = \frac{k_0^2|\bar{\varepsilon}|\mathbf{l}_i}{\mathbf{k}_i^2} \qquad (3.101)$$

Substituting equation (3.101) into (3.100) we find

$$[k_0^4|\bar{\varepsilon}| - (\mathbf{k}_i \cdot \bar{\varepsilon} \cdot \mathbf{k}_i)\mathbf{k}_i^2](\mathbf{l}_i \cdot \mathbf{H}_0) = 0 \qquad (3.102)$$

To satisfy condition (3.102), two possibilities arise: $\mathbf{H}_0 \cdot \mathbf{l}_i$ is zero or nonzero.

(a) $\mathbf{H}_0 \cdot \mathbf{l}_i = 0$, but $\mathbf{H}_0$ is also perpendicular on $\mathbf{k}$. Thus the direction of $\mathbf{H}_0$ is parallel to their cross product, i.e.,

$$\mathbf{h}_i \sim \mathbf{k}_i \times \mathbf{l}_i \qquad (no\ sum) \qquad (3.103)$$

finding the directions of $\mathbf{D}_0$ and $\mathbf{E}_0$:

$$\mathbf{d}_i \sim \frac{-\mathbf{k}_i \times \mathbf{h}_i}{\omega} = \frac{k_i^2\mathbf{l}_i}{\omega} \qquad (no\ sum) \qquad (3.104)$$

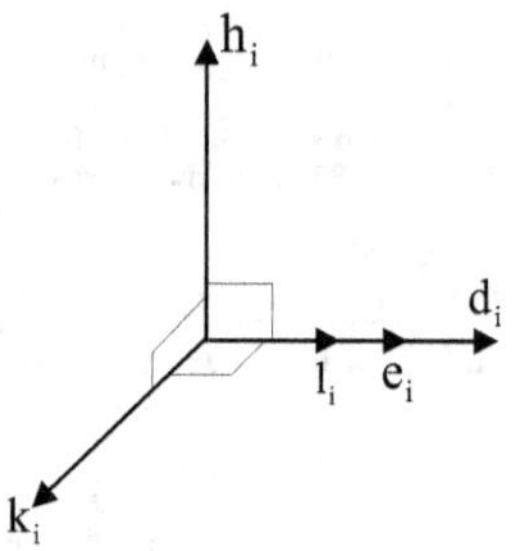

Figure 3.2.: Orientations of the Field Vectors when $k_0^2\overline{\varepsilon}$ is Singular.

and

$$\mathbf{e}_i \sim \frac{\overline{\varepsilon}^{\,-1} \cdot \mathbf{d}_i}{\varepsilon_0} = \omega\mu_0\varepsilon_i\overline{\varepsilon}^{\,-1} \cdot \mathbf{l}_i \qquad (no\ sum) \tag{3.105}$$

as shown in Figure 3.2 above, where $\overline{\varepsilon}^{\,-1} = \frac{\mathbf{l}_i\mathbf{l}_i}{\varepsilon_i}$, $k_i^2 = \omega^2\mu_0\varepsilon_0\varepsilon_i$

(b) $\mathbf{H}_0 \cdot \mathbf{l}_i \neq 0$. From equation (3.102) we have

$$k_0^4|\overline{\varepsilon}| - (\mathbf{k}_i \cdot \overline{\varepsilon} \cdot \mathbf{k}_i)\mathbf{k}_i^2 = 0 \tag{3.106}$$

which is the condition (3.61) denoting the coincidence of the given wave normal with the optic axis. In this case, the field vectors are limited only by the conditions implied by Maxwell's equations (3.7) and (3.8)

$$\mathbf{E}_0 \cdot \overline{\varepsilon} \cdot \widehat{\mathbf{k}} = 0 \qquad \mathbf{D}_0 \cdot \mathbf{H}_0 = 0 \qquad \widehat{\mathbf{k}} \cdot \mathbf{H}_0 = 0 \qquad \mathbf{H}_0 \cdot \mathbf{E}_0 = 0$$

3.4. Isonormal Waves

As shown by equation (3.62) for every direction of wave normal $\widehat{\mathbf{k}}$ there correspond two values of indices of refraction, n_+ and n_-, in crystals. The two values $\pm n_+^2$ or $\pm n_-^2$ are counted as once since the negative value evidently belongs to the opposite direction of wave normal $\widehat{\mathbf{k}}$. It follows that crystals permit two monochromatic plane waves with two different linear polarizations and two different phase velocities c/n_+ and c/n_- to propagate in any given direction of wave normal. Such waves having same $\widehat{\mathbf{k}}$ but different phase velocities are called *isonormal waves*.

According to equation (3.18), the two isonormal waves $\mathbf{H}_0^+$ and $\mathbf{H}_0^-$ satisfy

$$[(\widehat{\mathbf{k}} \times \overline{\mathbf{I}}) \cdot \overline{\varepsilon}^{\,-1} \cdot (\widehat{\mathbf{k}} \times \overline{\mathbf{I}})] \cdot \mathbf{H}_0^+ = \frac{-1}{n_+^2} \mathbf{H}_0^+ \tag{3.107}$$

$$[(\widehat{\mathbf{k}} \times \overline{\mathbf{I}}) \cdot \overline{\varepsilon}^{\,-1} \cdot (\widehat{\mathbf{k}} \times \overline{\mathbf{I}})] \cdot \mathbf{H}_0^- = \frac{-1}{n_-^2} \mathbf{H}_0^- \tag{3.108}$$

respectively. Dot-premultiplying equation (3.107) by $\mathbf{H}_0^-$ and equation (3.108) by $\mathbf{H}_0^+$, and then subtract. Noting that the matrices $\overline{\varepsilon}$ and $(\widehat{\mathbf{k}} \times \overline{\mathbf{I}}) \cdot \overline{\varepsilon}^{\,-1} \cdot (\widehat{\mathbf{k}} \times \overline{\mathbf{I}})$ are symmetric and $n_+^2 \neq n_-^2$, we obtain [12]

$$\mathbf{H}_0^+ \cdot \mathbf{H}_0^- = 0 \tag{3.109}$$

which means the the two isonormal waves are orthogonal. Also from Maxwell's equations

$$\omega\mu_0 \mathbf{H}_0^+ = k_+(\widehat{\mathbf{k}} \times \mathbf{E}_0^+) \tag{3.110}$$

$$\omega\mu_0 \mathbf{H}_0^- = k_-(\widehat{\mathbf{k}} \times \mathbf{E}_0^-) \tag{3.111}$$

$$\omega \mathbf{D}_0^+ = -k_+(\widehat{\mathbf{k}} \times \mathbf{H}_0^+) \tag{3.112}$$

$$\omega \mathbf{D}_0^- = -k_-(\widehat{\mathbf{k}} \times \mathbf{H}_0^-) \tag{3.113}$$

[12] Tensors $\overline{\varepsilon}^{\,-1} \cdot (\widehat{\mathbf{k}} \times \overline{\mathbf{I}})^2$, $(\widehat{\mathbf{k}} \times \overline{\mathbf{I}})^2 \cdot \overline{\varepsilon}^{\,-1}$, and $(\widehat{\mathbf{k}} \times \overline{\mathbf{I}}) \cdot \overline{\varepsilon}^{\,-1} \cdot (\widehat{\mathbf{k}} \times \overline{\mathbf{I}})$ are the products of tensors $\overline{\varepsilon}^{\,-1}$, $(\widehat{\mathbf{k}} \times \overline{\mathbf{I}})$, and $(\widehat{\mathbf{k}} \times \overline{\mathbf{I}})$ in cyclic order, where the trace and the determinant of the product of these three matrices in cyclic order are invariant.

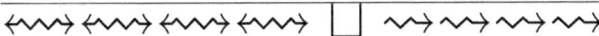

We conclude that

$$\hat{\mathbf{k}} \perp \mathbf{H}_0^+ \qquad \mathbf{H}_0^+ \perp \mathbf{H}_0^- \qquad \perp \mathbf{H}_0^- \perp \hat{\mathbf{k}} \tag{3.114}$$

$$\hat{\mathbf{k}} \perp \mathbf{D}_0^+ \qquad \mathbf{D}_0^+ \perp \mathbf{H}_0^+ \qquad \perp \mathbf{H}_0^+ \perp \hat{\mathbf{k}} \tag{3.115}$$

$$\hat{\mathbf{k}} \perp \mathbf{D}_0^- \qquad \mathbf{D}_0^- \perp \mathbf{H}_0^- \qquad \perp \mathbf{H}_0^- \perp \hat{\mathbf{k}} \tag{3.116}$$

Comparing equation (3.114) with equations (3.115) and (3.116)

$$\mathbf{D}_0^+ \parallel \mathbf{H}_0^- \qquad \mathbf{D}_0^- \parallel \mathbf{H}_0^+ \qquad \mathbf{D}_0^+ \parallel \mathbf{D}_0^- \tag{3.117}$$

From equations [(3.110), (3.111)] and (3.114)

$$\mathbf{E}_0^+ \parallel \mathbf{H}_0^+ \qquad \mathbf{E}_0^+ \parallel \mathbf{D}_0^- \tag{3.118}$$

$$\mathbf{E}_0^- \parallel \mathbf{H}_0^- \qquad \mathbf{E}_0^- \parallel \mathbf{D}_0^+ \tag{3.119}$$

Figure 3.3 below shows the orientations of field vectors of isonormal waves. The above relations are always valid except when the wave normal $\hat{\mathbf{k}}$ coincides with an optic axis. In the latter case $n_+ = n_-$, and the two isonormal waves become one.

3.5. Determination of Wave Vectors; The Booker Quartic

In an unbounded crystal as have been treated in the proceedings sections before, the dispersion equation alone didn't uniquely determine the wave vector. According to laws of reflection and refraction, we may represent any one of the wave vectors $\mathbf{k}_i$, $\mathbf{k}_r$, $\mathbf{k}_+$, $\mathbf{k}_-$ of the incident, the reflected and the two transmitted waves in the general form

$$\mathbf{k}_\alpha = \mathbf{b} + q_\alpha \hat{\mathbf{q}} \tag{3.120}$$

where the transcript α denotes i, r, +, -, and

$$\begin{aligned} \mathbf{b} &= \hat{\mathbf{q}} \times \mathbf{a} \\ \mathbf{a} &= \mathbf{k}_i \times \hat{\mathbf{q}} = \mathbf{b} \times \hat{\mathbf{q}} \end{aligned} \tag{3.121}$$

as discussed before.

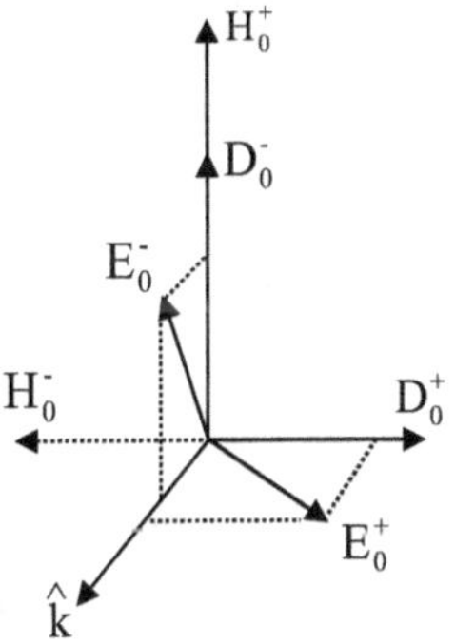

Figure 3.3.: Orientations of the Field Vectors of the Two Isonormal Waves.

To determine the wave vectors in an electrically anisotropic medium 2, we substitute equation (3.120) into the dispersion equation (3.34) and obtain

$$Aq_\alpha^4 + Bq_\alpha^3 + Cq_\alpha^2 + Dq_\alpha + F = 0 \tag{3.122}$$

where

$$\begin{aligned}
A &= \widehat{\mathbf{q}} \cdot \overline{\varepsilon} \cdot \widehat{\mathbf{q}} \\
B &= \mathbf{b} \cdot (\overline{\varepsilon} + \widetilde{\overline{\varepsilon}}) \cdot \widehat{\mathbf{q}} \\
C &= k_0^2 \widehat{\mathbf{q}} \cdot [\text{adj } \overline{\varepsilon} - (\text{adj } \overline{\varepsilon})_t \overline{\mathbf{I}}] \cdot \widehat{\mathbf{q}} + (\mathbf{b} \cdot \overline{\varepsilon} \cdot \mathbf{b}) + \mathbf{a}^2 (\widehat{\mathbf{q}} \cdot \overline{\varepsilon} \cdot \widehat{\mathbf{q}}) \\
D &= k_0^2 \mathbf{b} \cdot (\text{adj } \overline{\varepsilon} + \text{adj } \widetilde{\overline{\varepsilon}}) \cdot \widehat{\mathbf{q}} + \mathbf{a}^2 \mathbf{b} \cdot (\overline{\varepsilon} + \widetilde{\overline{\varepsilon}}) \cdot \widehat{\mathbf{q}} \\
F &= k_0^2 \mathbf{b} \cdot [\text{adj } \overline{\varepsilon} - (\text{adj } \overline{\varepsilon})_t \overline{\mathbf{I}}] \cdot \mathbf{b} + \mathbf{a}^2 (\mathbf{b} \cdot \overline{\varepsilon} \cdot \mathbf{b}) + k_0^4 |\overline{\varepsilon}|
\end{aligned} \tag{3.123}$$

In the case of a lossless crystal, $\overline{\varepsilon}$ is symmetric. Thus equation (3.123) becomes

$$\begin{aligned}
A &= \widehat{\mathbf{q}}\cdot\overline{\varepsilon}\cdot\widehat{\mathbf{q}} \\
B &= 2(\mathbf{b}\cdot\overline{\varepsilon}\cdot\widehat{\mathbf{q}}) \\
C &= k_0^2\widehat{\mathbf{q}}\cdot[\text{adj }\overline{\varepsilon}-(\text{adj }\overline{\varepsilon})_t\overline{\mathbf{I}}]\cdot\widehat{\mathbf{q}}+(\mathbf{b}\cdot\overline{\varepsilon}\cdot\mathbf{b})+\mathbf{a}^2(\widehat{\mathbf{q}}\cdot\overline{\varepsilon}\cdot\widehat{\mathbf{q}}) \qquad (3.124) \\
D &= 2k_0^2\mathbf{b}\cdot(\text{adj }\overline{\varepsilon})\cdot\widehat{\mathbf{q}}+2\mathbf{a}^2(\mathbf{b}\cdot\overline{\varepsilon}\cdot\widehat{\mathbf{q}}) \\
F &= k_0^2\mathbf{b}\cdot[\text{adj }\overline{\varepsilon}-(\text{adj }\overline{\varepsilon})_t\overline{\mathbf{I}}]\cdot\mathbf{b}+\mathbf{a}^2(\mathbf{b}\cdot\overline{\varepsilon}\cdot\mathbf{b})+k_0^4|\overline{\varepsilon}|
\end{aligned}$$

Equation (3.122), known as *Booker Quartic*, is a complete of fourth degree. It reduces to a biquadratic equation in the following special cases.

Normal Incidence
In this case $\mathbf{k}_\alpha = q_\alpha\widehat{\mathbf{q}}$ and equation (3.122) become

$$(\widehat{\mathbf{q}}\cdot\overline{\varepsilon}\cdot\widehat{\mathbf{q}})q_\alpha^4+k_0^2\widehat{\mathbf{q}}\cdot[\text{adj }\overline{\varepsilon}-(\text{adj }\overline{\varepsilon})_t\overline{\mathbf{I}}]\cdot\widehat{\mathbf{q}}q_\alpha^2+k_0^4=0 \qquad (3.125)$$

which is of the same form as the dispersion equation (3.34).

3.6. Reflection and Transmission of Waves at Plane Interface

Assuming that a uniform plane wave of arbitrary polarization is incident from an isotropic medium 1 on an anisotropic medium 2 as shown in Figure 3.4 below.

Decomposing the amplitude vectors of the incident and the reflected waves in medium 1 into components perpendicular and parallel to the plane of incidence

$$\begin{aligned}
\mathbf{E}_{0i} &= A_\perp\mathbf{a}+A_\parallel(\widehat{\mathbf{k}}_i\times\mathbf{a}) \qquad (3.126) \\
\mathbf{H}_{0i}=\frac{1}{\omega\mu_0}(\mathbf{k}_i\times\mathbf{E}_{0i}) &= \sqrt{\frac{\varepsilon_0\varepsilon_r}{\mu_0}}[A_\perp(\widehat{\mathbf{k}}_i\times\mathbf{a})-A_\parallel\mathbf{a}]
\end{aligned}$$

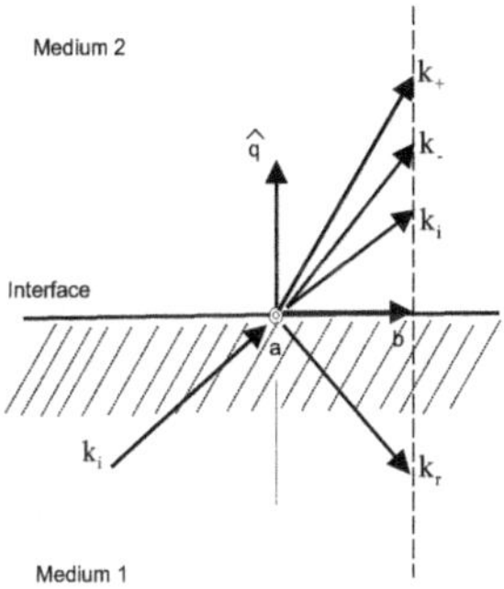

Figure 3.4.: Reflection and Transmission of the Waves at the Interface of an Isotropic-Anisotropic Medium. The Vector **a** is Directed out of the Paper. The Plane of Incidence Lies on the Paper.

and the reflected wave

$$\mathbf{E}_{0r} = B_\perp \mathbf{a} + B_\| (\widehat{\mathbf{k}}_r \times \mathbf{a}) \tag{3.127}$$

$$\mathbf{H}_{0r} = \frac{1}{\omega\mu_0}(\mathbf{k}_i \times \mathbf{E}_{0r}) = \sqrt{\frac{\varepsilon_0\varepsilon_r}{\mu_0}}[B_\perp(\widehat{\mathbf{k}}_r \times \mathbf{a}) - B_\| \mathbf{a}]$$

The subscripts + and - denote the two transmitted waves in medium 2. The two solutions q_+ and q_- out of four of Booker Quartic are selected such that they satisfy the condition which give the directions of energy propagation that are directed toward medium 2 and not toward medium 1. The wave vectors $\mathbf{k}_+$ and $\mathbf{k}_-$ are found from equation (3.120). Substituting these vectors separately into equations (3.84) and (3.105), we get the directions of the electric field intensities $\mathbf{e}_+$ and $\mathbf{e}_-$, and write the two transmitted waves as

$$\mathbf{E}_{0+} = C_+\mathbf{e}_+$$

$$\mathbf{H}_{0+} = \frac{1}{\omega\mu_0}(\mathbf{k}_+ \times \mathbf{E}_{0+}) = C_+\mathbf{h}_+ \tag{3.128}$$

where

$$\mathbf{h}_+ = \frac{1}{\omega\mu_0}(\mathbf{k}_+ \times \mathbf{e}_+)$$

and

$$\begin{aligned} \mathbf{E}_{0-} &= C_-\mathbf{e}_- \\ \mathbf{H}_{0-} &= \frac{1}{\omega\mu_0}(\mathbf{k}_- \times \mathbf{E}_{0-}) = C_-\mathbf{h}_- \end{aligned} \tag{3.129}$$

where

$$\mathbf{h}_- = \frac{1}{\omega\mu_0}(\mathbf{k}_- \times \mathbf{e}_-)$$

C_+ and C_- are arbitrary constants.

At the interface $\mathbf{r} \cdot \widehat{\mathbf{q}}=0$ and the field vectors satisfy the boundary conditions [13]

$$(\mathbf{E}_{0i} + \mathbf{E}_{0r} - \mathbf{E}_{0+} - \mathbf{E}_{0-}) \times \widehat{\mathbf{q}} = \mathbf{0}$$

or

$$\mathbf{E}_{0i} + \mathbf{E}_{0r} - \mathbf{E}_{0+} - \mathbf{E}_{0-} = \alpha\widehat{\mathbf{q}} \tag{3.130}$$

and

$$(\mathbf{H}_{0i} + \mathbf{H}_{0r} - \mathbf{H}_{0+} - \mathbf{H}_{0-}) \times \widehat{\mathbf{q}} = \mathbf{0}$$

or

$$\mathbf{H}_{0i} + \mathbf{H}_{0r} - \mathbf{H}_{0+} - \mathbf{H}_{0-} = \beta\widehat{\mathbf{q}} \tag{3.131}$$

where α and β are two constants.

Substituting equations (3.126) to (3.129) into equations (3.130) and (3.131) we obtain

$$\begin{aligned} (A_\perp + B_\perp)\mathbf{a} + A_\parallel(\widehat{\mathbf{k}}_i \times \mathbf{a}) + B_\parallel(\widehat{\mathbf{k}}_r \times \mathbf{a}) \\ - C_+\mathbf{e}_+ - C_-\mathbf{e}_- = \alpha\widehat{\mathbf{q}} \end{aligned} \tag{3.132}$$

[13] Regarding equation (1.27), $\mathbf{r}_v=0$ on the interface.

and

$$(A_{\|} + B_{\|})\mathbf{a} - A_{\perp}(\widehat{\mathbf{k}}_i \times \mathbf{a}) - B_{\perp}(\widehat{\mathbf{k}}_r \times \mathbf{a})$$
$$+ \sqrt{\frac{\mu_0}{\varepsilon_0\varepsilon_1}}C_+\mathbf{h}_+ + \sqrt{\frac{\mu_0}{\varepsilon_0\varepsilon_1}}C_-\mathbf{h}_- = -\sqrt{\frac{\mu_0}{\varepsilon_0\varepsilon_1}}\beta\widehat{\mathbf{q}} \tag{3.133}$$

taking the dot products of equations (3.132) and (3.133) respectively with vectors $\mathbf{a}$ and $\mathbf{b} = \widehat{\mathbf{q}} \times \mathbf{a}$, we get

$$A_{\perp} + B_{\perp} - \frac{\mathbf{a}\cdot\mathbf{e}_+}{\mathbf{a}^2}C_+ - \frac{\mathbf{a}\cdot\mathbf{e}_-}{\mathbf{a}^2}C_- = 0 \tag{3.134}$$

$$A_{\|} - B_{\|} - \frac{k_i(\mathbf{b}\cdot\mathbf{e}_+)}{q_i\mathbf{a}^2}C_+ - \frac{k_i(\mathbf{b}\cdot\mathbf{e}_-)}{q_i\mathbf{a}^2}C_- = 0 \tag{3.135}$$

$$A_{\|} + B_{\|} + \frac{\omega\mu_0(\mathbf{a}\cdot\mathbf{h}_+)}{k_i\mathbf{a}^2}C_+ + \frac{\omega\mu_0(\mathbf{a}\cdot\mathbf{h}_-)}{k_i\mathbf{a}^2}C_- = 0 \tag{3.136}$$

$$A_{\perp} - B_{\perp} - \frac{q_+(\mathbf{a}\cdot\mathbf{e}_+)}{q_i\mathbf{a}^2}C_+ - \frac{q_-(\mathbf{a}\cdot\mathbf{e}_-)}{q_i\mathbf{a}^2}C_- = 0 \tag{3.137}$$

In deriving equations (3.135) and (3.137), we have used

$$\mathbf{b}\cdot(\widehat{\mathbf{k}}_i \times \mathbf{a}) = \frac{q_i\mathbf{a}^2}{k_i} = -\mathbf{b}\cdot(\widehat{\mathbf{k}}_r \times \mathbf{a})$$
$$\mathbf{b}\cdot\mathbf{h}_{\pm} = \frac{q_{\pm}}{\omega\mu_0}(\mathbf{a}\cdot\mathbf{e}_{\pm}) \tag{3.138}$$

Eliminating $B_{\perp}$ or $A_{\perp}$ from equations (3.134) and (3.137), and $B_{\|}$ or $A_{\|}$ from equations (3.135) and (3.136), we have

$$
\begin{aligned}
& \frac{(q_i + q_+)(\mathbf{a}\cdot\mathbf{e}_+)}{2q_i\mathbf{a}^2}C_+ \\
+ & \frac{(q_i + q_-)(\mathbf{a}\cdot\mathbf{e}_-)}{2q_i\mathbf{a}^2}C_- \\
= & A_\perp \qquad (3.139)
\end{aligned}
$$

$$
\begin{aligned}
& \frac{(q_i - q_+)(\mathbf{a}\cdot\mathbf{e}_+)}{2q_i\mathbf{a}^2}C_+ \\
+ & \frac{(q_i - q_-)(\mathbf{a}\cdot\mathbf{e}_-)}{2q_i\mathbf{a}^2}C_- \\
= & B_\perp \qquad (3.140)
\end{aligned}
$$

$$
\begin{aligned}
& \frac{k_i^2(\mathbf{b}\cdot\mathbf{e}_+) - \omega\mu_0 q_i(\mathbf{a}\cdot\mathbf{h}_+)}{2k_i q_i\mathbf{a}^2}C_+ \\
+ & \frac{k_i^2(\mathbf{b}\cdot\mathbf{e}_-) - \omega\mu_0 q_i(\mathbf{a}\cdot\mathbf{h}_-)}{2k_i q_i\mathbf{a}^2}C_- \\
= & A_\parallel \qquad (3.141)
\end{aligned}
$$

$$
\begin{aligned}
- & \frac{k_i^2(\mathbf{b}\cdot\mathbf{e}_+) + \omega\mu_0 q_i(\mathbf{a}\cdot\mathbf{h}_+)}{2k_i q_i\mathbf{a}^2}C_+ \\
- & \frac{k_i^2(\mathbf{b}\cdot\mathbf{e}_-) + \omega\mu_0 q_i(\mathbf{a}\cdot\mathbf{h}_-)}{2k_i q_i\mathbf{a}^2}C_- \\
= & B_\parallel \qquad (3.142)
\end{aligned}
$$

To put the last four equations in a more symmetrical and compact form, we introduce vectors $\mathbf{N}_+$, $\mathbf{N}_-$, $\mathbf{F}_+$ and $\mathbf{F}_-$ defined by

$$
\begin{aligned}
\mathbf{N}_\pm &= \mathbf{k}_i^2\mathbf{e}_\pm + \omega\mu_0(\mathbf{k}_r \times \mathbf{h}_\pm) \qquad (3.143) \\
\mathbf{F}_\pm &= \mathbf{k}_i^2\mathbf{e}_\pm + \omega\mu_0(\mathbf{k}_i \times \mathbf{h}_\pm) \qquad (3.144)
\end{aligned}
$$

Or, using Maxwell's equations and after some simplification, we arrive at

$$\begin{aligned}\mathbf{N}_{\pm} &= (\mathbf{k}_i^2 - \mathbf{k}_r \cdot \mathbf{k}_{\pm})\mathbf{e}_{\pm} + (\mathbf{k}_r \cdot \mathbf{e}_{\pm})\mathbf{k}_{\pm} \\ &= q_i(q_i + q_{\pm})\mathbf{e}_{\pm} + (\mathbf{k}_r \cdot \mathbf{e}_{\pm})\mathbf{k}_{\pm} \end{aligned} \tag{3.145}$$

$$\begin{aligned}\mathbf{F}_{\pm} &= (\mathbf{k}_i^2 - \mathbf{k}_i \cdot \mathbf{k}_{\pm})\mathbf{e}_{\pm} + (\mathbf{k}_i \cdot \mathbf{e}_{\pm})\mathbf{k}_{\pm} \\ &= q_i(q_i - q_{\pm})\mathbf{e}_{\pm} + (\mathbf{k}_i \cdot \mathbf{e}_{\pm})\mathbf{k}_{\pm} \end{aligned} \tag{3.146}$$

Thus

$$\begin{aligned}\mathbf{a} \cdot \mathbf{N}_{\pm} &= q_i(q_i + q_{\pm})(\mathbf{a} \cdot \mathbf{e}_{\pm}) \\ \mathbf{a} \cdot \mathbf{F}_{\pm} &= q_i(q_i - q_{\pm})(\mathbf{a} \cdot \mathbf{e}_{\pm}) \\ \mathbf{b} \cdot \mathbf{N}_{\pm} &= \mathbf{k}_i^2(\mathbf{b} \cdot \mathbf{e}_{\pm}) - \omega\mu_0 q_i(\mathbf{a} \cdot \mathbf{h}_{\pm}) \\ &= (\mathbf{k}_i^2 + q_i q_{\pm})(\mathbf{b} \cdot \mathbf{e}_{\pm}) - q_i \mathbf{a}^2(\widehat{\mathbf{q}} \cdot \mathbf{e}_{\pm}) \\ \mathbf{b} \cdot \mathbf{F}_{\pm} &= \mathbf{k}_i^2(\mathbf{b} \cdot \mathbf{e}_{\pm}) + \omega\mu_0 q_i(\mathbf{a} \cdot \mathbf{h}_{\pm}) \\ &= (\mathbf{k}_i^2 - q_i q_{\pm})(\mathbf{b} \cdot \mathbf{e}_{\pm}) + q_i \mathbf{a}^2(\widehat{\mathbf{q}} \cdot \mathbf{e}_{\pm}) \end{aligned} \tag{3.147}$$

Substituting equations (3.147) into equations (3.139) and (3.141) before finally putting the result into a matrix

$$\frac{1}{2q_i\mathbf{a}^2}\begin{bmatrix} \frac{1}{q_i}(\mathbf{a} \cdot \mathbf{N}_+) & \frac{1}{q_i}(\mathbf{a} \cdot \mathbf{N}_-) \\ \frac{1}{k_i}(\mathbf{b} \cdot \mathbf{N}_+) & \frac{1}{k_i}(\mathbf{b} \cdot \mathbf{N}_-) \end{bmatrix} * \begin{bmatrix} C_+ \\ C_- \end{bmatrix} = \begin{bmatrix} A_{\perp} \\ A_{\parallel} \end{bmatrix} \tag{3.148}$$

Multiplying both sides of equation (3.148) by the inverse coefficient matrix, we obtain

$$\begin{bmatrix} C_+ \\ C_- \end{bmatrix} = \begin{bmatrix} T_{11} & T_{12} \\ T_{21} & T_{22} \end{bmatrix} \begin{bmatrix} A_{\perp} \\ A_{\parallel} \end{bmatrix} \tag{3.149}$$

where the transmission coefficients are given by

$$\begin{aligned} T_{11} &= \frac{2q_i^2\mathbf{a}^2}{\Delta}(\mathbf{b}\cdot\mathbf{N}_-) \\ T_{12} &= -\frac{2q_ik_i\mathbf{a}^2}{\Delta}(\mathbf{a}\cdot\mathbf{N}_-) \\ T_{21} &= -\frac{2q_i^2\mathbf{a}^2}{\Delta}(\mathbf{b}\cdot\mathbf{N}_+) \\ T_{22} &= \frac{2q_ik_i\mathbf{a}^2}{\Delta}(\mathbf{a}\cdot\mathbf{N}_+) \end{aligned} \tag{3.150}$$

and

$$\Delta = (\mathbf{a}\cdot\mathbf{N}_+)(\mathbf{b}\cdot\mathbf{N}_-) - (\mathbf{a}\cdot\mathbf{N}_-)(\mathbf{b}\cdot\mathbf{N}_+) \tag{3.151}$$

Similarly, using equations (3.147) we may write equations (3.140) and (3.142) in the following matrix form

$$\frac{1}{2q_i\mathbf{a}^2}\begin{bmatrix} \frac{1}{q_i}(\mathbf{a}\cdot\mathbf{F}_+) & \frac{1}{q_i}(\mathbf{a}\cdot\mathbf{F}_-) \\ \frac{-1}{k_i}(\mathbf{b}\cdot\mathbf{F}_+) & \frac{-1}{k_i}(\mathbf{b}\cdot\mathbf{F}_-) \end{bmatrix} * \begin{bmatrix} C_+ \\ C_- \end{bmatrix} = \begin{bmatrix} B_\perp \\ B_\parallel \end{bmatrix} \tag{3.152}$$

Substituting the result of equation (3.149) into equation (3.152) and after some simplification, we obtain

$$\begin{bmatrix} B_\perp \\ B_\parallel \end{bmatrix} = \begin{bmatrix} \Gamma_{11} & \Gamma_{12} \\ \Gamma_{21} & \Gamma_{22} \end{bmatrix}\begin{bmatrix} A_\perp \\ A_\parallel \end{bmatrix} \tag{3.153}$$

where the reflection coefficients are given by

$$
\begin{aligned}
\Gamma_{11} &= \frac{1}{\Delta}[(\mathbf{a}\cdot\mathbf{F}_{+})(\mathbf{b}\cdot\mathbf{N}_{-})-(\mathbf{a}\cdot\mathbf{F}_{-})(\mathbf{b}\cdot\mathbf{N}_{+})] \\
\Gamma_{12} &= -\frac{k_i}{q_i\Delta}[(\mathbf{a}\cdot\mathbf{F}_{+})(\mathbf{a}\cdot\mathbf{N}_{-})-(\mathbf{a}\cdot\mathbf{F}_{-})(\mathbf{a}\cdot\mathbf{N}_{+})] \\
\Gamma_{21} &= -\frac{q_i}{k_i\Delta}[(\mathbf{b}\cdot\mathbf{F}_{+})(\mathbf{b}\cdot\mathbf{N}_{-})-(\mathbf{b}\cdot\mathbf{F}_{-})(\mathbf{b}\cdot\mathbf{N}_{+})] \\
\Gamma_{22} &= \frac{1}{\Delta}[(\mathbf{b}\cdot\mathbf{F}_{+})(\mathbf{a}\cdot\mathbf{N}_{-})-(\mathbf{b}\cdot\mathbf{F}_{-})(\mathbf{a}\cdot\mathbf{N}_{+})]
\end{aligned}
\qquad (3.154)
$$

3.7. Plane Waves in Lossless, Nonmagnetic and Homogeneous Anisotropic Uniaxial Media

3.7.1. Wave Matrix of a Uniaxial Medium

For an orthogonal coordinate system, the dielectric tensor of a uniaxial medium takes the matrix form

$$
\overline{\varepsilon} = \begin{bmatrix} \varepsilon_{\perp} & 0 & 0 \\ 0 & \varepsilon_{\perp} & 0 \\ 0 & 0 & \varepsilon_{\parallel} \end{bmatrix} \qquad (3.155)
$$

where

$$
\overline{\varepsilon} = \varepsilon_{\perp}\overline{\mathbf{I}} + (\varepsilon_{\parallel} - \varepsilon_{\perp})\widehat{\mathbf{c}}\widehat{\mathbf{c}} \qquad (3.156)
$$

where $\widehat{\mathbf{c}}$ is a unit eigenvector of $\overline{\varepsilon}$ corresponding to the nonrepeated eigenvalue $\varepsilon_{\parallel}$, and $\varepsilon_{\perp}$ (repeated) is the other eigenvalue of $\overline{\varepsilon}$. From equation (3.156) and (3.59), we have

$$
\begin{aligned}
|\overline{\varepsilon}| &= \varepsilon_{\perp}^{2}\varepsilon_{\parallel} \\
\mathrm{adj}\ \overline{\varepsilon} &= \varepsilon_{\perp}[\varepsilon_{\parallel}\overline{\mathbf{I}} + (\varepsilon_{\parallel} - \varepsilon_{\perp})\widehat{\mathbf{c}}\widehat{\mathbf{c}}] \\
(\mathrm{adj}\ \overline{\varepsilon})_t\overline{\mathbf{I}} - \mathrm{adj}\ \overline{\varepsilon} &= \varepsilon_{\perp}(\varepsilon_{\parallel}\overline{\mathbf{I}} + \overline{\varepsilon})
\end{aligned}
\qquad (3.157)
$$

Substitution of equation (3.156) into equation (3.12) gives

$$\overline{\mathbf{W}}_u(\mathbf{k}) \cdot \mathbf{E}_0 = \mathbf{0} \tag{3.158}$$

where $\overline{\mathbf{W}}_u$, the wave matrix of a uniaxial medium, is a function of $\mathbf{k}$ and is given by

$$\overline{\mathbf{W}}_u(\mathbf{k}) = (k_0^2\varepsilon_\perp - \mathbf{k}^2)\overline{\mathbf{I}} + \mathbf{k}\mathbf{k} + k_0^2(\varepsilon_\parallel - \varepsilon_\perp)\hat{\mathbf{c}}\hat{\mathbf{c}} \tag{3.159}$$

From equation (3.158) we see that the multiplication of any solution of the homogeneous equation by a constant yields another solution, equation (3.158) therefore uniquely determines only the direction of $\mathbf{E}_0$ and not its magnitude.

Using

$$\begin{array}{rcl} \det\,(\alpha\overline{\mathbf{I}}+ & \mathbf{A}\mathbf{B} & +\mathbf{C}\mathbf{D}) \\ & = & \\ \alpha[\alpha(\alpha+\mathbf{A}\cdot\mathbf{B}+\mathbf{C}\cdot\mathbf{D}) & + & (\mathbf{B}\times\mathbf{D})\cdot(\mathbf{A}\times\mathbf{C})] \end{array} \tag{3.160}$$

$$\begin{array}{rcl} (\mathbf{B}\times\mathbf{D}) & \cdot & (\mathbf{A}\times\mathbf{C}) \\ & = & \\ \mathbf{C}\cdot\mathbf{D}\mathbf{A}\cdot\mathbf{B} & - & \mathbf{B}\cdot\mathbf{C}\mathbf{A}\cdot\mathbf{D} \end{array} \tag{3.161}$$

$$\begin{array}{rcl} \mathrm{adj}\,(\alpha\overline{\mathbf{I}}+ & \mathbf{A}\mathbf{B} & +\mathbf{C}\mathbf{D}) \\ & = & \\ \alpha[(\alpha+\mathbf{A}\cdot\mathbf{B}+\mathbf{C}\cdot\mathbf{D})\overline{\mathbf{I}}- & \mathbf{A}\mathbf{B} & -\mathbf{C}\mathbf{D}]+(\mathbf{B}\times\mathbf{D})(\mathbf{A}\times\mathbf{C}) \end{array} \tag{3.162}$$

we get

$$|\overline{\mathbf{W}}_u(\mathbf{k})| = k_0^2(\mathbf{k}^2 - k_0^2\varepsilon_\perp)[(\mathbf{k}\cdot\overline{\varepsilon}\cdot\mathbf{k}) - k_0^2\varepsilon_\perp\varepsilon_\parallel] \tag{3.163}$$

$$\begin{aligned} \mathrm{adj}\,\overline{\mathbf{W}}_u(\mathbf{k}) &= (k_0^2\varepsilon_\perp - \mathbf{k}^2)[k_0^2\varepsilon_\parallel\overline{\mathbf{I}} - \mathbf{k}\mathbf{k} - k_0^2(\varepsilon_\parallel - \varepsilon_\perp)\hat{\mathbf{c}}\hat{\mathbf{c}}] \\ &\quad + k_0^2(\varepsilon_\parallel - \varepsilon_\perp)(\mathbf{k}\times\hat{\mathbf{c}})(\mathbf{k}\times\hat{\mathbf{c}}) \end{aligned} \tag{3.164}$$

A nonzero solution $\mathbf{E}_0$ of equation (3.158) exists provided that $|\overline{\mathbf{W}}_u(\mathbf{k})|=0$.

Regarding [1], for those values of $\mathbf{k}$ where we obtain nonzero solutions, we have a planar matrix $\overline{\mathbf{W}}_u$, thus:
$|\overline{\mathbf{W}}_u|=0 \implies \mathrm{adj}\,\overline{\mathbf{W}}_u \neq 0$ and

$$\mathbf{E}_0 = [\mathrm{adj}\,\overline{\mathbf{W}}_u(\mathbf{k})] \cdot \mathbf{u} \tag{3.165}$$

3.7.2. Determination of the Directions of the Field Vectors

By setting the two terms of a product in equation (3.163) to zero, we obtain the dispersion equations

$$\mathbf{k}^2 = k_0^2 \varepsilon_\perp \qquad (3.166)$$

$$\mathbf{k} \cdot \overline{\varepsilon} \cdot \mathbf{k} = k_0^2 \varepsilon_\perp \varepsilon_\parallel \qquad (3.167)$$

Equation (3.166) does not depend on the direction of wave normal and the corresponding wave is called the *ordinary wave*, where

$$k_+ = k_0 \sqrt{\varepsilon_\perp} \qquad (3.168)$$

On the other hand, equation (3.167) does depend on the direction of wave normal, and the corresponding wave is called the *extraordinary wave*, where

$$k_- = k_0 \sqrt{\frac{\varepsilon_\perp \varepsilon_\parallel}{\widehat{\mathbf{k}}_- \cdot \overline{\varepsilon} \cdot \widehat{\mathbf{k}}_-}} \qquad (3.169)$$

Substituting equation (3.156) into equation (3.169), we obtain

$$k_- = k_0 \sqrt{\frac{\varepsilon_\perp \varepsilon_\parallel}{\varepsilon_\perp (\widehat{\mathbf{k}}_- \times \widehat{\mathbf{c}})^2 + \varepsilon_\parallel (\widehat{\mathbf{k}}_- \cdot \widehat{\mathbf{c}})^2}} \qquad (3.170)$$

If $\widehat{\mathbf{k}}_-$ coincides with the vector $\widehat{\mathbf{c}}$, then $(\widehat{\mathbf{k}}_- \cdot \widehat{\mathbf{c}})^2 = 1$, $(\widehat{\mathbf{k}}_- \times \widehat{\mathbf{c}})^2 = 0$ and

$$k_- = k_0 \sqrt{\varepsilon_\perp} = k_+$$

Which means that the wave numbers and the phase velocities of the two isonormal waves coincide only when wave number $\widehat{\mathbf{k}}_-$ is along $\widehat{\mathbf{c}}$. Therefore $\widehat{\mathbf{c}}$ is the optic axis. And since there is only one optic axis of the medium, we call such a medium as *uniaxial*.

For the ordinary wave equation (3.164) becomes

$$\text{adj } \overline{\mathbf{W}}_u(\mathbf{k}) = k_0^2 (\varepsilon_\parallel - \varepsilon_\perp)(\mathbf{k}_+ \times \widehat{\mathbf{c}})(\mathbf{k}_+ \times \widehat{\mathbf{c}}) \qquad (3.171)$$

According to equation (3.165), we may choose the direction of $\mathbf{E}_0$ as

$$\mathbf{e}_+ \sim \mathbf{k}_+ \times \widehat{\mathbf{c}} \qquad (3.172)$$

and the directions of the remaining field vectors are to be determined from Maxwell's equations (see Figure 3.5 below)

$$\mathbf{h}_+ \sim \frac{1}{\omega\mu_0}[\mathbf{k}_+ \times (\mathbf{k}_+ \times \widehat{\mathbf{c}})] \quad (3.173)$$

$$\mathbf{b}_+ \sim \frac{1}{\omega}[\mathbf{k}_+ \times (\mathbf{k}_+ \times \widehat{\mathbf{c}})] \quad (3.174)$$

$$\mathbf{d}_+ \sim \varepsilon_0\varepsilon_\perp(\mathbf{k}_+ \times \widehat{\mathbf{c}}) \quad (3.175)$$

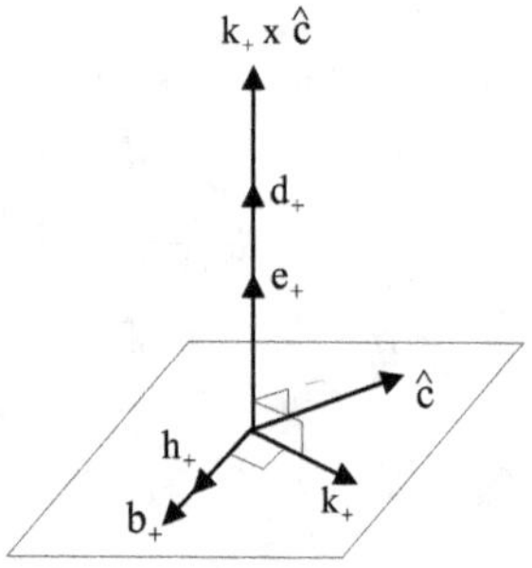

Figure 3.5.: Orientations of the Field Vectors of the Ordinary Wave in a Uniaxial Medium.

For the extraordinary wave equation (3.164) becomes

$$\text{adj } \overline{\mathbf{W}}_u(\mathbf{k}_-) = (k_0^2\varepsilon_\perp - \mathbf{k}_-^2)[k_0^2\varepsilon_\parallel\overline{\mathbf{I}} - \mathbf{k}_-\mathbf{k}_- - k_0^2(\varepsilon_\parallel - \varepsilon_\perp)\widehat{\mathbf{c}}\widehat{\mathbf{c}}] + k_0^2(\varepsilon_\parallel - \varepsilon_\perp)(\mathbf{k}_- \times \widehat{\mathbf{c}})(\mathbf{k}_- \times \widehat{\mathbf{c}}) \quad (3.176)$$

choosing an arbitrary vector such as $\mathbf{u} = \widehat{\mathbf{c}}$ in equation (3.165), would give the direction of $\mathbf{E}_0$ for the extraordinary wave

$$\mathbf{e}_- \sim k_0\varepsilon_\perp\widehat{\mathbf{c}} - (\frac{\mathbf{k}_-}{k_0}\cdot\widehat{\mathbf{c}})\mathbf{k}_- \tag{3.177}$$

and from Maxwell's equations, we have the directions of the remaining field vectors (see Figure 3.6 below)

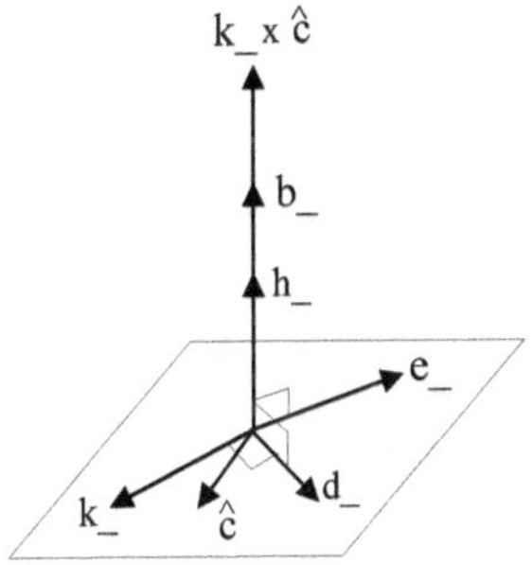

Figure 3.6.: Orientations of the Field Vectors of the Extraordinary Wave in a Uniaxial Medium.

$$\mathbf{h}_- \sim \omega\varepsilon_0\varepsilon_\perp(\frac{\mathbf{k}_-}{k_0}\times\widehat{\mathbf{c}}) \tag{3.178}$$

$$\mathbf{b}_- \sim \omega\mu_0\varepsilon_0\varepsilon_\perp(\frac{\mathbf{k}_-}{k_0}\times\widehat{\mathbf{c}}) \tag{3.179}$$

$$\mathbf{d}_- \sim -\varepsilon_0\varepsilon_\perp[\mathbf{k}_-\times(\frac{\mathbf{k}_-}{k_0}\times\widehat{\mathbf{c}})] \tag{3.180}$$

3.7.3. Determination of Wave Vectors at Isotropic-Uniaxial Interface

Assuming a nonmagnetic properties for both medium 1 (isotropic) and medium 2 (uniaxial) we determine both wave vectors.
Having

$$\mathbf{k}_+ = \mathbf{b} + q_+\widehat{\mathbf{q}} \tag{3.181}$$

as the transmitted ordinary wave, we substitute equation (3.181) into the dispersion equation (3.166) and find

$$q_+ = \sqrt{k_0^2\varepsilon_\perp - \mathbf{a}^2} \tag{3.182}$$

Similarly, substituting the wave vector

$$\mathbf{k}_- = \mathbf{b} + q_-\widehat{\mathbf{q}} \tag{3.183}$$

of the transmitted extraordinary wave into the dispersion equation (3.167), we obtain

$$\begin{aligned}(\widehat{\mathbf{q}}\cdot\overline{\varepsilon}\cdot\widehat{\mathbf{q}})q_-^2 + 2(\mathbf{b}\cdot\overline{\varepsilon}\cdot\widehat{\mathbf{q}})q_- \quad &+ \quad (\mathbf{b}\cdot\overline{\varepsilon}\cdot\mathbf{b}) - k_0^2\varepsilon_\perp\varepsilon_\parallel \\ &= \\ &0\end{aligned} \tag{3.184}$$

which is a quadratic equation in q_-. q_- must be chosen out of the solutions of the former equation (refer to [1]) so that $\mathbf{s}\cdot\widehat{\mathbf{q}} > 0$. And this condition corresponds to $\mathbf{k}_-\cdot\overline{\varepsilon}\cdot\widehat{\mathbf{q}} > 0$ or

$$\begin{aligned}(\mathbf{b} + q_-\widehat{\mathbf{q}})\cdot\overline{\varepsilon}\cdot\widehat{\mathbf{q}} &> 0 \\ \mathbf{b}\cdot\overline{\varepsilon}\cdot\widehat{\mathbf{q}} &> -q_-\widehat{\mathbf{q}}\cdot\overline{\varepsilon}\cdot\widehat{\mathbf{q}} \\ q_- &> -\frac{\mathbf{b}\cdot\overline{\varepsilon}\cdot\widehat{\mathbf{q}}}{\widehat{\mathbf{q}}\cdot\overline{\varepsilon}\cdot\widehat{\mathbf{q}}}\end{aligned} \tag{3.185}$$

From equations (3.184) and (3.185) we obtain

$$q_- = \frac{-(\mathbf{b}\cdot\overline{\varepsilon}\cdot\widehat{\mathbf{q}}) + \zeta}{\widehat{\mathbf{q}}\cdot\overline{\varepsilon}\cdot\widehat{\mathbf{q}}} \tag{3.186}$$

where

$$\zeta = [(\mathbf{b}\cdot\overline{\varepsilon}\cdot\widehat{\mathbf{q}})^2 - (\widehat{\mathbf{q}}\cdot\overline{\varepsilon}\cdot\widehat{\mathbf{q}})(\mathbf{b}\cdot\overline{\varepsilon}\cdot\mathbf{b} - k_0^2\varepsilon_\perp\varepsilon_\parallel)]^{1/2}$$

The expression inside the square root can be simplified further since

$$(\mathbf{b}\cdot\overline{\varepsilon}\cdot\widehat{\mathbf{q}})^2-(\widehat{\mathbf{q}}\cdot\overline{\varepsilon}\cdot\widehat{\mathbf{q}})(\mathbf{b}\cdot\overline{\varepsilon}\cdot\mathbf{b})=(\widehat{\mathbf{q}}\times\mathbf{b})\cdot[(\overline{\varepsilon}\cdot\mathbf{b})\times(\overline{\varepsilon}\cdot\widehat{\mathbf{q}})]$$

and having

$$(\overline{\varepsilon}\cdot\mathbf{b})\times(\overline{\varepsilon}\cdot\widehat{\mathbf{q}})=(\text{adj }\overline{\varepsilon})\cdot(\mathbf{b}\times\widehat{\mathbf{q}})$$

Thus

$$\zeta=[k_0^2\varepsilon_\perp\varepsilon_\parallel(\widehat{\mathbf{q}}\cdot\overline{\varepsilon}\cdot\widehat{\mathbf{q}})-\mathbf{a}\cdot(\text{adj }\overline{\varepsilon})\cdot\mathbf{a}]^{1/2}$$

where $\mathbf{b}\times\widehat{\mathbf{q}}=\mathbf{a}$ is used. Substituting equation (3.186) into equation (3.183) and noting that

$$\begin{aligned}\mathbf{b}(\widehat{\mathbf{q}}\cdot\overline{\varepsilon}\cdot\widehat{\mathbf{q}})-\widehat{\mathbf{q}}(\mathbf{b}\cdot\overline{\varepsilon}\cdot\widehat{\mathbf{q}}) &= (\mathbf{b}\widehat{\mathbf{q}}-\widehat{\mathbf{q}}\mathbf{b})\cdot(\overline{\varepsilon}\cdot\widehat{\mathbf{q}})\\ &= [(\widehat{\mathbf{q}}\times\mathbf{b})\times\overline{\mathbf{I}}]\cdot(\overline{\varepsilon}\cdot\widehat{\mathbf{q}})\\ &= (\overline{\varepsilon}\cdot\widehat{\mathbf{q}})\times\mathbf{a}\end{aligned}$$

we finally obtain

$$\mathbf{k}_-=\frac{(\overline{\varepsilon}\cdot\widehat{\mathbf{q}})\times\mathbf{a}+\zeta\widehat{\mathbf{q}}}{\widehat{\mathbf{q}}\cdot\overline{\varepsilon}\cdot\widehat{\mathbf{q}}}\tag{3.187}$$

and in medium 1 we have (see Figure 3.7 below)

$$\mathbf{k}_i=\mathbf{b}+q_i\widehat{\mathbf{q}}\tag{3.188}$$

$$\mathbf{k}_r=\mathbf{b}-q_i\widehat{\mathbf{q}}\tag{3.189}$$

$$q_i=\sqrt{k_0^2\varepsilon_1-\mathbf{a}^2}\tag{3.190}$$

3.7.4. Reflection and Transmission Coefficients

To calculate the reflection and transmission coefficients we use the equations (3.126) and (3.127) of medium 1, and for the transmitted waves we have

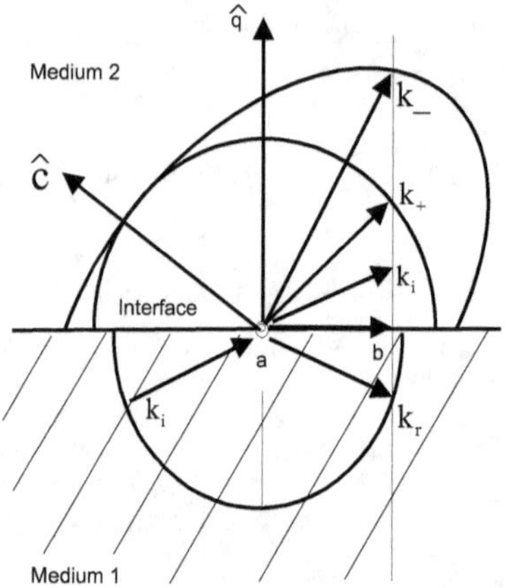

Figure 3.7.: Geometrical determination of a wave vectors at the interface of an isotropic and a uniaxial medium. Vector **a** is directed out of the page.

Ordinary wave:

$$\begin{aligned} \mathbf{E}_{0+} &= C_+ \mathbf{e}_+ \\ \mathbf{H}_{0+} &= C_+ \mathbf{h}_+ \end{aligned} \tag{3.191}$$

where

$$\begin{aligned} \mathbf{e}_+ &\sim \mathbf{k}_+ \times \widehat{\mathbf{c}} \\ \mathbf{h}_+ &\sim \frac{1}{\omega\mu_0}(\mathbf{k}_+ \times \mathbf{e}_+) = \frac{1}{\omega\mu_0}[\mathbf{k}_+ \times (\mathbf{k}_+ \times \widehat{\mathbf{c}})] \end{aligned} \tag{3.192}$$

and

$$\begin{aligned} \mathbf{k}_+ &= \mathbf{b} + q_+ \widehat{\mathbf{q}} \\ q_+ &= \sqrt{k_0^2 \varepsilon_\perp - \mathbf{a}^2} \\ k_+ &= k_0 \sqrt{\varepsilon_\perp} \end{aligned} \tag{3.193}$$

Extraordinary wave:

$$\begin{aligned} \mathbf{E}_{0-} &= C_- \mathbf{e}_- \\ \mathbf{H}_{0-} &= C_- \mathbf{h}_- \end{aligned} \tag{3.194}$$

where

$$\begin{aligned} \mathbf{e}_- &\sim k_0 \varepsilon_\perp \widehat{\mathbf{c}} - (\frac{\mathbf{k}_-}{k_0} \cdot \widehat{\mathbf{c}}) \mathbf{k}_- \\ \mathbf{h}_- &\sim \omega \varepsilon_0 \varepsilon_\perp (\frac{\mathbf{k}_-}{k_0} \times \widehat{\mathbf{c}}) \end{aligned} \tag{3.195}$$

Equations (3.169) and (3.187) give the extraordinary wave vector. The constants $A_\perp$ and $A_\parallel$ are assumed to be given, and the constants $B_\perp$, $B_\parallel$, C_+ and C_- are to be determined. Figure 3.8 below shows the orientation of the optic axis with respect to the plane of incidence and the interface.

We find that

$$\begin{aligned}
\mathbf{a}\cdot\mathbf{e}_{+} &= -\mathbf{k}_{+}\cdot(\mathbf{a}\times\widehat{\mathbf{c}}) \\
\mathbf{a}\cdot\mathbf{e}_{-} &= k_0\varepsilon_{\perp}(\mathbf{a}\cdot\widehat{\mathbf{c}}) \\
\mathbf{a}\cdot\mathbf{h}_{+} &= -\omega\varepsilon_0\varepsilon_{\perp}(\mathbf{a}\cdot\widehat{\mathbf{c}}) \\
\mathbf{a}\cdot\mathbf{h}_{-} &= -\omega\varepsilon_0\varepsilon_{\perp}[\frac{\mathbf{k}_{-}}{k_0}\cdot(\mathbf{a}\times\widehat{\mathbf{c}})] \\
\mathbf{b}\cdot\mathbf{e}_{+} &= \mathbf{b}\times\mathbf{k}_{+}\cdot\widehat{\mathbf{c}} = \mathbf{b}\times[\mathbf{b}+q_{+}\widehat{\mathbf{q}}]\cdot\widehat{\mathbf{c}} \\
&= q_{+}(\mathbf{a}\cdot\widehat{\mathbf{c}}) \\
\mathbf{b}\cdot\mathbf{e}_{-} &= \mathbf{b}\cdot(k_0\varepsilon_{\perp}\widehat{\mathbf{c}} - (\frac{\mathbf{k}_{-}}{k_0}\cdot\widehat{\mathbf{c}})\mathbf{k}_{-}) \\
&= \mathbf{b}\cdot(k_0\varepsilon_{\perp}\widehat{\mathbf{c}} - (\frac{[\mathbf{b}+q_{-}\widehat{\mathbf{q}}]}{k_0}\cdot\widehat{\mathbf{c}})[\mathbf{b}+q_{-}\widehat{\mathbf{q}}]) \\
&= k_0\varepsilon_{\perp}(\mathbf{b}\cdot\widehat{\mathbf{c}}) - \frac{\mathbf{a}^2}{k_0}(\mathbf{b}\cdot\widehat{\mathbf{c}}) - \frac{q_{-}}{k_0}\mathbf{a}^2(\widehat{\mathbf{q}}\cdot\widehat{\mathbf{c}})
\end{aligned} \tag{3.196}$$

Substitution of equation (3.196) into equation (3.147) yields

$$
\begin{aligned}
\mathbf{a}\cdot\mathbf{N}_+ &= q_i(q_i+q_+)(-\mathbf{k}_+\cdot(\mathbf{a}\times\widehat{\mathbf{c}}))\\
\mathbf{a}\cdot\mathbf{N}_- &= q_i(q_i+q_-)(k_0\varepsilon_\perp(\mathbf{a}\cdot\widehat{\mathbf{c}}))\\
\mathbf{a}\cdot\mathbf{F}_+ &= q_i(q_i-q_+)(-\mathbf{k}_+\cdot(\mathbf{a}\times\widehat{\mathbf{c}}))\\
\mathbf{a}\cdot\mathbf{F}_- &= q_i(q_i-q_-)(k_0\varepsilon_\perp(\mathbf{a}\cdot\widehat{\mathbf{c}}))\\
\mathbf{b}\cdot\mathbf{N}_+ &= \mathbf{k}_i^2(q_+(\mathbf{a}\cdot\widehat{\mathbf{c}}))-\omega\mu_0 q_i(-\omega\varepsilon_0\varepsilon_\perp(\mathbf{a}\cdot\widehat{\mathbf{c}}))\\
&= (\mathbf{k}_i^2+q_iq_+)(q_+(\mathbf{a}\cdot\widehat{\mathbf{c}}))-q_i\mathbf{a}^2(\widehat{\mathbf{q}}\cdot(\mathbf{k}_+\times\widehat{\mathbf{c}}))\\
\mathbf{b}\cdot\mathbf{N}_- &= \mathbf{k}_i^2(k_0\varepsilon_\perp(\mathbf{b}\cdot\widehat{\mathbf{c}})-\frac{\mathbf{a}^2}{k_0}(\mathbf{b}\cdot\widehat{\mathbf{c}})-\frac{q_-}{k_0}\mathbf{a}^2(\widehat{\mathbf{q}}\cdot\widehat{\mathbf{c}}))\\
&\quad-\omega\mu_0 q_i(-\omega\varepsilon_0\varepsilon_\perp[\frac{\mathbf{k}_-}{k_0}\cdot(\mathbf{a}\times\widehat{\mathbf{c}})])\\
&= (\mathbf{k}_i^2+q_iq_-)(k_0\varepsilon_\perp(\mathbf{b}\cdot\widehat{\mathbf{c}})-\frac{\mathbf{a}^2}{k_0}(\mathbf{b}\cdot\widehat{\mathbf{c}})-\frac{q_-}{k_0}\mathbf{a}^2(\widehat{\mathbf{q}}\cdot\widehat{\mathbf{c}}))\\
&\quad-q_i\mathbf{a}^2(\widehat{\mathbf{q}}\cdot(k_0\varepsilon_\perp\widehat{\mathbf{c}}-(\frac{\mathbf{k}_-}{k_0}\cdot\widehat{\mathbf{c}})\mathbf{k}_-))\\
\mathbf{b}\cdot\mathbf{F}_+ &= \mathbf{k}_i^2(q_+(\mathbf{a}\cdot\widehat{\mathbf{c}}))+\omega\mu_0 q_i(-\omega\varepsilon_0\varepsilon_\perp(\mathbf{a}\cdot\widehat{\mathbf{c}}))\\
&= (\mathbf{k}_i^2-q_iq_+)(q_+(\mathbf{a}\cdot\widehat{\mathbf{c}}))+q_i\mathbf{a}^2(\widehat{\mathbf{q}}\cdot(\mathbf{k}_+\times\widehat{\mathbf{c}}))\\
\mathbf{b}\cdot\mathbf{F}_- &= \mathbf{k}_i^2(k_0\varepsilon_\perp(\mathbf{b}\cdot\widehat{\mathbf{c}})-\frac{\mathbf{a}^2}{k_0}(\mathbf{b}\cdot\widehat{\mathbf{c}})\\
&\quad-\frac{q_-}{k_0}\mathbf{a}^2(\widehat{\mathbf{q}}\cdot\widehat{\mathbf{c}}))+\omega\mu_0 q_i(-\omega\varepsilon_0\varepsilon_\perp[\frac{\mathbf{k}_-}{k_0}\cdot(\mathbf{a}\times\widehat{\mathbf{c}})])\\
&= (\mathbf{k}_i^2-q_iq_-)(k_0\varepsilon_\perp(\mathbf{b}\cdot\widehat{\mathbf{c}})-\frac{\mathbf{a}^2}{k_0}(\mathbf{b}\cdot\widehat{\mathbf{c}})-\frac{q_-}{k_0}\mathbf{a}^2(\widehat{\mathbf{q}}\cdot\widehat{\mathbf{c}}))\\
&\quad+q_i\mathbf{a}^2(\widehat{\mathbf{q}}\cdot(k_0\varepsilon_\perp\widehat{\mathbf{c}}-(\frac{\mathbf{k}_-}{k_0}\cdot\widehat{\mathbf{c}})\mathbf{k}_-))
\end{aligned}
\tag{3.197}
$$

With more simplified form [14]

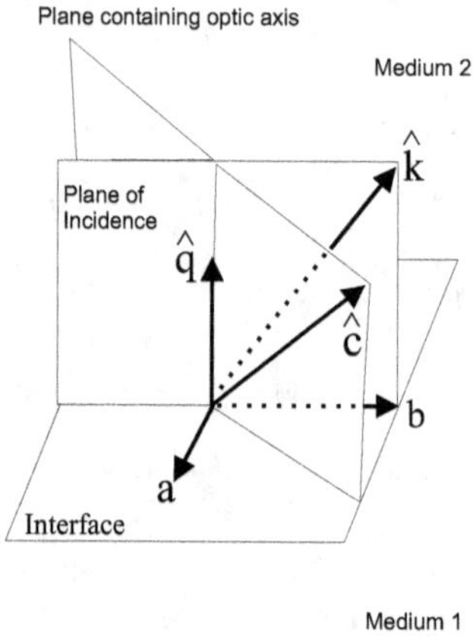

Figure 3.8.: Orientation of the Plane of Incidence, the Interface and the Plane Containing the Optic Axis.

[14] $k_i^2 = k_0^2 \varepsilon_1$

$$
\begin{aligned}
\mathbf{a}\cdot\mathbf{N}_+ &= -q_i(q_i+q_+)\mathbf{k}_+\cdot(\mathbf{a}\times\widehat{\mathbf{c}}) \\
\mathbf{a}\cdot\mathbf{N}_- &= k_0\varepsilon_\perp q_i(q_i+q_-)(\mathbf{a}\cdot\widehat{\mathbf{c}}) \\
\mathbf{b}\cdot\mathbf{N}_+ &= (\varepsilon_1 q_+ + \varepsilon_\perp q_i)k_0^2(\mathbf{a}\cdot\widehat{\mathbf{c}}) \\
\mathbf{b}\cdot\mathbf{N}_- &= \frac{k_i^2}{k_0}[q_+^2(\mathbf{b}\cdot\widehat{\mathbf{c}}) - q_-\mathbf{a}^2(\widehat{\mathbf{q}}\cdot\widehat{\mathbf{c}})] + k_0^2\varepsilon_\perp q_i[\frac{\mathbf{k}_-}{k_0}\cdot(\mathbf{a}\times\widehat{\mathbf{c}})] \\
\mathbf{a}\cdot\mathbf{F}_+ &= q_i(q_+-q_i)[\mathbf{k}_+\cdot(\mathbf{a}\times\widehat{\mathbf{c}})] \\
\mathbf{a}\cdot\mathbf{F}_- &= k_0\varepsilon_\perp q_i(q_i-q_-)(\mathbf{a}\cdot\widehat{\mathbf{c}}) \\
\mathbf{b}\cdot\mathbf{F}_+ &= (\varepsilon_1 q_+ - \varepsilon_\perp q_i)k_0^2(\mathbf{a}\cdot\widehat{\mathbf{c}}) \\
\mathbf{b}\cdot\mathbf{F}_- &= \frac{k_i^2}{k_0}[q_+^2(\mathbf{b}\cdot\widehat{\mathbf{c}}) - q_-\mathbf{a}^2(\widehat{\mathbf{q}}\cdot\widehat{\mathbf{c}})] - k_0^2\varepsilon_\perp q_i[\frac{\mathbf{k}_-}{k_0}\cdot(\mathbf{a}\times\widehat{\mathbf{c}})]
\end{aligned}
\tag{3.198}
$$

Substituting equation (3.198) into equation (3.150), we find the transmission coefficients

$$
\begin{aligned}
T_{11} &= \tfrac{M_{11}}{\Delta_1} \qquad T_{12} = \tfrac{M_{12}}{\Delta_1} \\
T_{21} &= \tfrac{M_{21}}{\Delta_1} \qquad T_{22} = \tfrac{M_{22}}{\Delta_1}
\end{aligned}
\tag{3.199}
$$

where

$$\begin{aligned}
M_{11} &= \frac{-2q_i\mathbf{a}^2(X+Y)}{\mathbf{k}_+\cdot(\mathbf{a}\times\widehat{\mathbf{c}})} \\
M_{21} &= \frac{2q_i\mathbf{a}^2(U+Z)}{k_0\varepsilon_\perp(\mathbf{a}\cdot\widehat{\mathbf{c}})} \\
&= 2k_0q_i\mathbf{a}^2(k_i^2q_+ + k_0^2\varepsilon_\perp q_i)(\mathbf{a}\cdot\widehat{\mathbf{c}}) \\
M_{12} &= 2k_0^2k_iq_i\mathbf{a}^2\varepsilon_\perp(q_i+q_-)(\mathbf{a}\cdot\widehat{\mathbf{c}}) \\
M_{22} &= 2k_0k_iq_i\mathbf{a}^2(q_i+q_+)[\mathbf{k}_+\cdot(\mathbf{a}\times\widehat{\mathbf{c}})]
\end{aligned} \tag{3.200}$$

and

$$\begin{aligned}
\Delta_1 &= (q_i+q_+)(X+Y)+(q_i+q_-)(U+Z) \\
X &= k_0^2q_i\varepsilon_\perp[\mathbf{k}_+\cdot(\mathbf{a}\times\widehat{\mathbf{c}})][\mathbf{k}_-\cdot(\mathbf{a}\times\widehat{\mathbf{c}})] \\
Y &= k_i^2[\mathbf{k}_+\cdot(\mathbf{a}\times\widehat{\mathbf{c}})][q_+^2(\mathbf{b}\cdot\widehat{\mathbf{c}})-q_-\mathbf{a}^2(\widehat{\mathbf{q}}\cdot\widehat{\mathbf{c}})] \\
U &= k_0^2k_i^2\varepsilon_\perp q_+(\mathbf{a}\cdot\widehat{\mathbf{c}})^2 \\
Z &= k_0^4\varepsilon_\perp^2q_i(\mathbf{a}\cdot\widehat{\mathbf{c}})^2
\end{aligned} \tag{3.201}$$

Similarly, substituting equation (3.198) into equation (3.154), we obtain the reflection coefficients

$$\begin{aligned}
\Gamma_{11} &= \frac{(q_i-q_+)(X+Y)+(q_i-q_-)(U+Z)}{\Delta_1} \\
\Gamma_{12} &= \frac{2(q_+-q_-)(V-L)}{\Delta_1}
\end{aligned} \tag{3.202}$$

$$\begin{aligned}
\Gamma_{21} &= \frac{2(q_--q_+)(V+L)}{\Delta_1} \\
\Gamma_{22} &= \frac{(q_i+q_+)(X-Y)+(q_i+q_-)(Z-U)}{\Delta_1}
\end{aligned} \tag{3.203}$$

where

$$V = k_0^2 \varepsilon_\perp k_i q_i q_+ (\mathbf{a} \cdot \widehat{\mathbf{c}})(\mathbf{b} \cdot \widehat{\mathbf{c}})$$
$$L = k_0^2 \varepsilon_\perp k_i q_i \mathbf{a}^2 (\mathbf{a} \cdot \widehat{\mathbf{c}})(\widehat{\mathbf{q}} \cdot \widehat{\mathbf{c}}) \tag{3.204}$$

and

$$V - L = k_0^2 \varepsilon_\perp k_i q_i (\mathbf{a} \cdot \widehat{\mathbf{c}})[\mathbf{k}_+ \cdot (\mathbf{a} \times \widehat{\mathbf{c}})] \tag{3.205}$$

$$V + L = k_0^2 \varepsilon_\perp k_i q_i (\mathbf{a} \cdot \widehat{\mathbf{c}})[q_+ (\mathbf{b} \cdot \widehat{\mathbf{c}}) - \mathbf{a}^2 (\widehat{\mathbf{q}} \cdot \widehat{\mathbf{c}})] \tag{3.206}$$

Normal Incidence

In case of normal incidence, former formulas are no longer valid because the concept of plane of incidence loses its meaning. Equally meaningless is decomposing the field vectors into components perpendicular and parallel to the plane of incidence. In this case, the wave vectors take form

$$\mathbf{k}_i = k_i \widehat{\mathbf{q}} = -\mathbf{k}_r, \qquad \mathbf{k}_+ = k_+ \widehat{\mathbf{q}}, \qquad \mathbf{k}_- = k_- \widehat{\mathbf{q}} \tag{3.207}$$

where

$$k_i = k_0 \sqrt{\varepsilon_1}, \quad k_+ = k_0 \sqrt{\varepsilon_\perp}, \quad k_- = k_0 \sqrt{\frac{\varepsilon_\perp \varepsilon_\|}{\widehat{\mathbf{q}} \cdot \overline{\varepsilon} \cdot \widehat{\mathbf{q}}}} \tag{3.208}$$

We may treat the plane formed by vectors $\widehat{\mathbf{c}}$ and $\widehat{\mathbf{q}}$ as though it were the plane of incidence. Henceforth the subscripts $\perp$ and $\|$ will be used in this sense. We decompose the field vectors of the incident and the reflected waves into components perpendicular and parallel to the plane formed by vectors $\widehat{\mathbf{c}}$ and $\widehat{\mathbf{q}}$ as shown in Figure 3.9 below.

Thus, the incident wave becomes

$$\mathbf{E}_{0i} = A_\perp (\widehat{\mathbf{q}} \times \widehat{\mathbf{c}}) + A_\| [\widehat{\mathbf{q}} \times (\widehat{\mathbf{q}} \times \widehat{\mathbf{c}})]$$
$$\mathbf{H}_{0i} = \frac{1}{\omega \mu_0} (\mathbf{k}_i \times \mathbf{E}_{0i}) = \sqrt{\frac{\varepsilon_0 \varepsilon_1}{\mu_0}} \{A_\perp [\widehat{\mathbf{q}} \times (\widehat{\mathbf{q}} \times \widehat{\mathbf{c}})] - A_\| (\widehat{\mathbf{q}} \times \widehat{\mathbf{c}})\} \tag{3.209}$$

The reflected wave takes the form

$$\mathbf{E}_{0r} = B_\perp (\widehat{\mathbf{q}} \times \widehat{\mathbf{c}}) + B_\| [\widehat{\mathbf{q}} \times (\widehat{\mathbf{q}} \times \widehat{\mathbf{c}})]$$
$$\mathbf{H}_{0r} = \frac{1}{\omega \mu_0} (\mathbf{k}_r \times \mathbf{E}_{0r}) = \sqrt{\frac{\varepsilon_0 \varepsilon_1}{\mu_0}} \{-B_\perp [\widehat{\mathbf{q}} \times (\widehat{\mathbf{q}} \times \widehat{\mathbf{c}})] + B_\| (\widehat{\mathbf{q}} \times \widehat{\mathbf{c}})\} \tag{3.210}$$

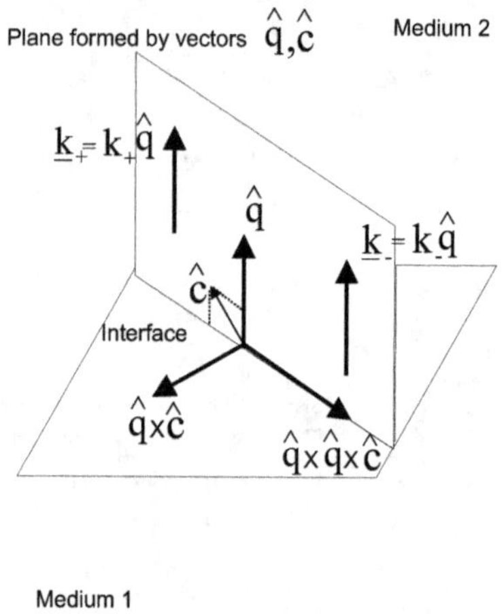

Figure 3.9.: Orientations of the Interface and the Plane Formed by the Vectors $\widehat{\mathbf{q}}$ and $\widehat{\mathbf{c}}$ in the Case of Normal Incidence.

And the transmitted waves, according to equations (3.192) and (3.195), are

Ordinary wave:

$$\begin{aligned} \mathbf{E}_{0+} &= C_+(\widehat{\mathbf{q}} \times \widehat{\mathbf{c}}) \\ \mathbf{H}_{0+} &= \frac{1}{\omega\mu_0}(\mathbf{k}_+ \times \mathbf{E}_{0+}) = \sqrt{\frac{\varepsilon_0\varepsilon_\perp}{\mu_0}}C_+[\widehat{\mathbf{q}} \times (\widehat{\mathbf{q}} \times \widehat{\mathbf{c}})] \end{aligned} \tag{3.211}$$

Extraordinary wave:

$$\begin{aligned} \mathbf{E}_{0-} &= C_-[\varepsilon_\perp\widehat{\mathbf{c}} - \frac{k_-^2}{k_0^2}(\widehat{\mathbf{q}} \cdot \widehat{\mathbf{c}})\widehat{\mathbf{q}}] \\ \mathbf{H}_{0-} &= \frac{1}{\omega\mu_0}(\mathbf{k}_- \times \mathbf{E}_{0-}) = \frac{1}{\omega\mu_0}\varepsilon_\perp k_- C_-(\widehat{\mathbf{q}} \times \widehat{\mathbf{c}}) \end{aligned} \tag{3.212}$$

Substituting equations (3.202) to (3.212) into the boundary conditions (3.130) and (2.40), we obtain

$$\begin{aligned} A_\perp \quad & (\widehat{\mathbf{q}} \times \widehat{\mathbf{c}}) + A_\parallel [\widehat{\mathbf{q}} \times (\widehat{\mathbf{q}} \times \widehat{\mathbf{c}})] + B_\perp (\widehat{\mathbf{q}} \times \widehat{\mathbf{c}}) + B_\parallel [\widehat{\mathbf{q}} \times (\widehat{\mathbf{q}} \times \widehat{\mathbf{c}})] \\ & -C_+ (\widehat{\mathbf{q}} \times \widehat{\mathbf{c}}) - C_- [\varepsilon_\perp \widehat{\mathbf{c}} - (\widehat{\mathbf{q}} \cdot \widehat{\mathbf{c}}) \widehat{\mathbf{q}}] = \alpha \widehat{\mathbf{q}} \end{aligned} \tag{3.213}$$

$$\begin{aligned} \sqrt{\frac{\varepsilon_0 \varepsilon_1}{\mu_0}} \{A_\perp [\widehat{\mathbf{q}} \times (\widehat{\mathbf{q}} \times \widehat{\mathbf{c}})] \quad & - \quad A_\parallel (\widehat{\mathbf{q}} \times \widehat{\mathbf{c}})\} \\ & + \quad \sqrt{\frac{\varepsilon_0 \varepsilon_1}{\mu_0}} \{-B_\perp [\widehat{\mathbf{q}} \times (\widehat{\mathbf{q}} \times \widehat{\mathbf{c}})] + B_\parallel (\widehat{\mathbf{q}} \times \widehat{\mathbf{c}})\} \\ - \sqrt{\frac{\varepsilon_0 \varepsilon_1}{\mu_0}} C_+ [\widehat{\mathbf{q}} \times (\widehat{\mathbf{q}} \times \widehat{\mathbf{c}})] \quad & - \quad \frac{1}{\omega \mu_0} \varepsilon_\perp k_- C_- (\widehat{\mathbf{q}} \times \widehat{\mathbf{c}}) = \beta \widehat{\mathbf{q}} \end{aligned} \tag{3.214}$$

Cross pre-multiplying equation (3.213) by $\widehat{\mathbf{q}}$ we get [15]

$$\begin{aligned} (A_\perp + B_\perp - C_+)[\widehat{\mathbf{q}} \times (\widehat{\mathbf{q}} \times \widehat{\mathbf{c}})] \quad & - \quad (A_\parallel + B_\parallel + C_- \ \varepsilon_\perp)(\widehat{\mathbf{q}} \times \widehat{\mathbf{c}}) \\ & = \\ & \mathbf{0} \end{aligned} \tag{3.215}$$

And solving for equation (2.40)

$$\begin{aligned} (A_\perp - B_\perp - \sqrt{\frac{\varepsilon_\perp}{\varepsilon_1}} C_+)[\widehat{\mathbf{q}} \times (\widehat{\mathbf{q}} \times \widehat{\mathbf{c}})] \quad & + \\ (B_\parallel - A_\parallel - \frac{\varepsilon_\perp}{\omega \sqrt{\varepsilon_0 \mu_0 \varepsilon_1}} k_- C_-)(\widehat{\mathbf{q}} \times \widehat{\mathbf{c}}) \quad & = \quad \mathbf{0} \end{aligned} \tag{3.216}$$

Since $\widehat{\mathbf{q}} \times \widehat{\mathbf{c}}$ and $\widehat{\mathbf{q}} \times (\widehat{\mathbf{q}} \times \widehat{\mathbf{c}})$ are two linearly independent vectors, it follows that

$$\begin{aligned} A_\perp + B_\perp - C_+ = 0 \\ A_\perp - B_\perp - \sqrt{\frac{\varepsilon_\perp}{\varepsilon_1}} C_+ = 0 \end{aligned} \tag{3.217}$$

[15] where $\widehat{\mathbf{q}} \times [\widehat{\mathbf{q}} \times (\widehat{\mathbf{q}} \times \widehat{\mathbf{c}})] = (\widehat{\mathbf{q}} \times \widehat{\mathbf{q}}) \overset{=0}{(\widehat{\mathbf{q}}} \cdot \widehat{\mathbf{c}}) - (\widehat{\mathbf{q}} \times \widehat{\mathbf{c}})(\widehat{\mathbf{q}} \cdot \widehat{\mathbf{q}}) = -(\widehat{\mathbf{q}} \times \widehat{\mathbf{c}})$

and

$$A_{\parallel} + B_{\parallel} + C_{-}\ \varepsilon_{\perp} = 0$$

$$A_{\parallel} - B_{\parallel} + \frac{\varepsilon_{\perp}}{\omega\sqrt{\varepsilon_0\mu_0\varepsilon_1}} k_{-} C_{-} = 0 \tag{3.218}$$

which can easily be solved; yielding

$$\frac{C_{+}}{A_{\perp}} = \frac{2k_i}{k_i + k_{+}} \qquad \frac{C_{-}}{A_{\parallel}} = \frac{-2k_i}{\varepsilon_{\perp}(k_i + k_{-})} \tag{3.219}$$

$$\frac{B_{\perp}}{A_{\perp}} = \frac{k_i - k_{+}}{k_i + k_{+}} \qquad \frac{B_{\parallel}}{A_{\parallel}} = \frac{k_i - k_{-}}{k_i + k_{-}} \tag{3.220}$$

However, we should note that in special case when the wave normals $\widehat{\mathbf{k}}_{+}$ and $\widehat{\mathbf{k}}_{-}$ of the two transmitted waves coincide with the optic axis $\widehat{\mathbf{c}}$, formulas (3.199) and (3.203) again lose their validity. Since in this case $k_{+}^2 = k_{-}^2 = k_0^2\varepsilon_{\perp} = k_t^2$ and $\mathbf{k}_{+} = \mathbf{k}_{-} = k_t\widehat{\mathbf{c}} = \mathbf{k}_t$, it follows that $\mathbf{e}_{+}$ and $\mathbf{e}_{-}$ become zero. But the condition $\mathbf{k}_t \cdot \mathbf{D}_0 = 0$ implies $\mathbf{c} \cdot \mathbf{E}_0 = 0$. Thus electric and magnetic field vectors of the transmitted wave are perpendicular to the wave normal $\widehat{\mathbf{k}}_t = \widehat{\mathbf{c}}$ and may have any directions in the plane perpendicular to $\widehat{\mathbf{c}}$. Analogous to the case of isotropic media, we may decompose the transmitted wave into components perpendicular and parallel to the plane of incidence. We use the results of equations (2.76), (2.79), (2.80) and (2.83) and note that $\mathbf{k}_t = k_0\sqrt{\varepsilon_{\perp}}\ \widehat{\mathbf{c}}$. Thus

$$\frac{B_{\perp}}{A_{\perp}} = \frac{\mathbf{a}\cdot(\widehat{\mathbf{c}} \times \mathbf{k}_i)}{\mathbf{a}\cdot(\mathbf{k}_r \times \widehat{\mathbf{c}})} \tag{3.221}$$

$$\frac{B_{\parallel}}{A_{\parallel}} = \frac{(\widehat{\mathbf{k}}_r\cdot\widehat{\mathbf{c}})\ [\mathbf{a}\cdot(\widehat{\mathbf{c}} \times \mathbf{k}_i)]}{(\widehat{\mathbf{k}}_i\cdot\widehat{\mathbf{c}})\ [\mathbf{a}\cdot(\mathbf{k}_r \times \widehat{\mathbf{c}})]} \tag{3.222}$$

$$\frac{C_{\perp}}{A_{\parallel}} = \frac{\mathbf{a}\cdot(\mathbf{k}_r \times \mathbf{k}_i)}{k_0\sqrt{\varepsilon_{\perp}}\mathbf{a}\cdot(\mathbf{k}_r \times \widehat{\mathbf{c}})} \tag{3.223}$$

$$\frac{C_{\parallel}}{A_{\parallel}} = \frac{\mathbf{a}\cdot(\mathbf{k}_r \times \mathbf{k}_i)}{(\mathbf{k}_i\cdot\widehat{\mathbf{c}})k_0\sqrt{\varepsilon_{\perp}}\mathbf{a}\cdot(\mathbf{k}_r \times \widehat{\mathbf{c}})} \tag{3.224}$$

3.7.5. Some Special Cases

Optic Axis Parallel to the Interface

When the optic axis of a uniaxial crystal is parallel to the interface but is arbitrary oriented with respect to the plane of incidence, we have the condition $\widehat{\mathbf{q}} \cdot \widehat{\mathbf{c}} = 0$. Then

$$\begin{aligned} \mathbf{k}_{\pm} \cdot (\mathbf{a} \cdot \widehat{\mathbf{c}}) &= q_{\pm}(\mathbf{b} \cdot \widehat{\mathbf{c}}) \\ \mathbf{b} \cdot \overline{\varepsilon} \cdot \widehat{\mathbf{q}} &= 0 \\ \widehat{\mathbf{q}} \cdot \overline{\varepsilon} \cdot \widehat{\mathbf{q}} &= \varepsilon_{\perp} \end{aligned} \tag{3.225}$$

Equations (3.201) and (3.204) are reduced to

$$\begin{aligned} X \pm Y &= q_{+}(\mathbf{b} \cdot \widehat{\mathbf{c}})^2 (k_0^2 \varepsilon_{\perp} q_i q_{-} \pm k_i^2 q_{+}^2) \\ Z \pm U &= k_0^2 \varepsilon_{\perp} (\mathbf{a} \cdot \widehat{\mathbf{c}})^2 (k_0^2 \varepsilon_{\perp} q_i \pm k_i^2 q_{+}) \\ L &= 0 \end{aligned} \tag{3.226}$$

and

$$(\mathbf{b} \cdot \widehat{\mathbf{c}})^2 = (\mathbf{a} \times \widehat{\mathbf{c}})^2 = \mathbf{a}^2 - (\mathbf{a} \cdot \widehat{\mathbf{c}})^2$$

We may write equation (3.186) as

$$q_{-} = \frac{1}{\varepsilon_{\perp}} \sqrt{\varepsilon_{\perp} \varepsilon_{\parallel} (k_0^2 \varepsilon_{\perp} - \mathbf{a}^2) - \varepsilon_{\perp} (\varepsilon_{\perp} - \varepsilon_{\parallel})(\mathbf{a} \cdot \widehat{\mathbf{c}})^2} \tag{3.227}$$

Substituting equations (3.225) and (3.226) into equations (3.199) and (3.202), we obtain, respectively, the transmission coefficients

$$\begin{aligned} T_{11} &= \frac{-2q_i \mathbf{a}^2 (\mathbf{b} \cdot \widehat{\mathbf{c}})(k_0^2 \varepsilon_{\perp} q_i q_{-} + k_i^2 q_{+}^2)}{\Delta_1} \\ T_{12} &= \frac{2k_0^2 \varepsilon_{\perp} k_i q_i \mathbf{a}^2 (\mathbf{a} \cdot \widehat{\mathbf{c}})(q_i + q_{-})}{\Delta_1} \\ T_{21} &= \frac{2q_i \mathbf{a}^2 (\mathbf{a} \cdot \widehat{\mathbf{c}})(k_i^2 q_{+} + k_0^2 \varepsilon_{\perp} q_i)}{\Delta_1} \\ T_{22} &= \frac{2k_i q_i q_{+} \mathbf{a}^2 (\mathbf{b} \cdot \widehat{\mathbf{c}})(q_i + q_{+})}{\Delta_1} \end{aligned} \tag{3.228}$$

and the reflection coefficients

$$\Gamma_{11} = \frac{(q_i - q_+)(X+Y) + (q_i - q_-)(U+Z)}{\Delta_1}$$

$$\Gamma_{12} = \frac{2(q_+ - q_-)(k_0^2\varepsilon_\perp k_i q_i q_+ (\mathbf{a}\cdot\widehat{\mathbf{c}})(\mathbf{b}\cdot\widehat{\mathbf{c}}))}{\Delta_1} \tag{3.229}$$

$$\Gamma_{21} = \frac{2(q_- - q_+)(k_0^2\varepsilon_\perp k_i q_i q_+ (\mathbf{a}\cdot\widehat{\mathbf{c}})(\mathbf{b}\cdot\widehat{\mathbf{c}}))}{\Delta_1} \tag{3.230}$$

$$\Gamma_{22} = \frac{(q_i + q_+)(X-Y) + (q_i + q_-)(Z-U)}{\Delta_1}$$

where

$$\Delta_1 = (q_i + q_+)(X+Y) + (q_i + q_-)(U+Z)$$

We have the following two cases:

(1) *Optic Axis Parallel to the Plane of Incidence*

In this case $\widehat{\mathbf{c}}$ is parallel to $\mathbf{b}$; thus $\mathbf{b}\cdot\widehat{\mathbf{c}} = |\mathbf{a}|$ and $\mathbf{a}\cdot\widehat{\mathbf{c}} = 0$. We have then

$$\begin{aligned} q_- &= \sqrt{\frac{\varepsilon_\parallel}{\varepsilon_\perp}}\sqrt{k_0^2\varepsilon_\perp - \mathbf{a}^2} = \sqrt{\frac{\varepsilon_\parallel}{\varepsilon_\perp}}\,q_+ \\ U &= Z = V = L = 0 \\ \Delta_1 &= q_+(q_i + q_+)\mathbf{a}^2(k_0^2\varepsilon_\perp q_i q_- + k_i^2 q_+^2) \end{aligned} \tag{3.231}$$

and the transmission and reflection coefficient become [16]

$$\begin{aligned} T_{11} &= \frac{-2q_i|\mathbf{a}|}{q_+(q_i+q_+)} \\ &= \frac{-2\sin\theta_i\cos\theta_i}{\cos\theta_i\sqrt{\varepsilon_\perp - \sin^2\theta_i} - (\varepsilon_\perp - \sin^2\theta_i)} \\ T_{22} &= \frac{2k_0 k_i q_i |\mathbf{a}|}{q_+(k_0^2\sqrt{\varepsilon_\perp\varepsilon_\parallel}\,q_i + k_i^2 q_+)} \\ &= \frac{2\sin\theta_i\cos\theta_i}{\sqrt{\varepsilon_\perp - \sin^2\theta_i}\,\sqrt{\varepsilon_\perp\varepsilon_\parallel}\cos\theta_i + \varepsilon_\perp - \sin^2\theta_i} \end{aligned} \tag{3.232}$$

$$T_{21} = T_{12} = 0$$

[16] all solutions here are taken in free space where $\varepsilon_1 = \mu_1 = 1$

and

$$
\begin{aligned}
\Gamma_{11} &= \frac{q_i - q_+}{q_i + q_+} \\
&= \frac{\cos\theta_i - \sqrt{\varepsilon_\perp - \sin^2\theta_i}}{\cos\theta_i + \sqrt{\varepsilon_\perp - \sin^2\theta_i}} \\
\Gamma_{22} &= \frac{(\sqrt{\varepsilon_\perp \varepsilon_\parallel}\, q_i - \varepsilon_1 q_+)}{(\sqrt{\varepsilon_\perp \varepsilon_\parallel}\, q_i + \varepsilon_1 q_+)} \\
&= \frac{\sqrt{\varepsilon_\perp \varepsilon_\parallel}\cos\theta_i - \sqrt{\varepsilon_\perp - \sin^2\theta_i}}{\sqrt{\varepsilon_\perp \varepsilon_\parallel}\cos\theta_i + \sqrt{\varepsilon_\perp - \sin^2\theta_i}}
\end{aligned}
\tag{3.233}
$$

$$
\Gamma_{12} = \Gamma_{21} = 0 \tag{3.234}
$$

Figure 3.10 below shows the reflection coefficient varying with the angle of incidence and the relative permittivity. We notice that the reflection coefficient goes to zero where Brewster's angle takes place, while the slope of the inclination within higher incident angle values increases sharply if the parallel permittivity is to be increased when maintaining the perpendicular permittivity's value.

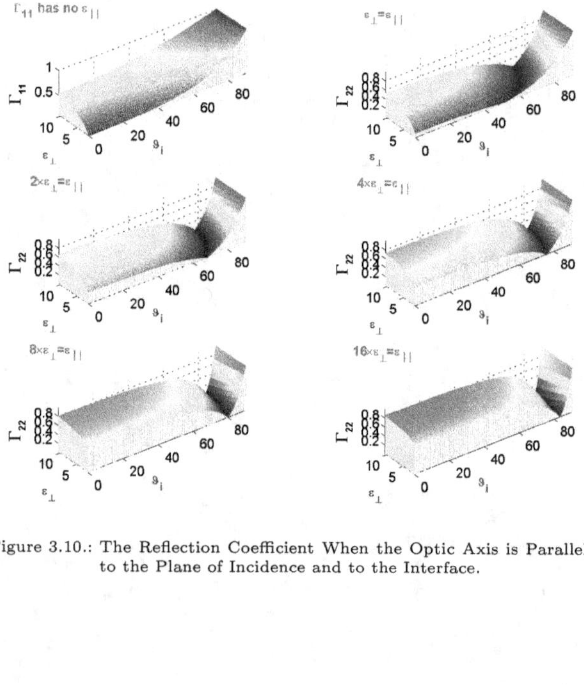

Figure 3.10.: The Reflection Coefficient When the Optic Axis is Parallel to the Plane of Incidence and to the Interface.

And with a top view (see below)

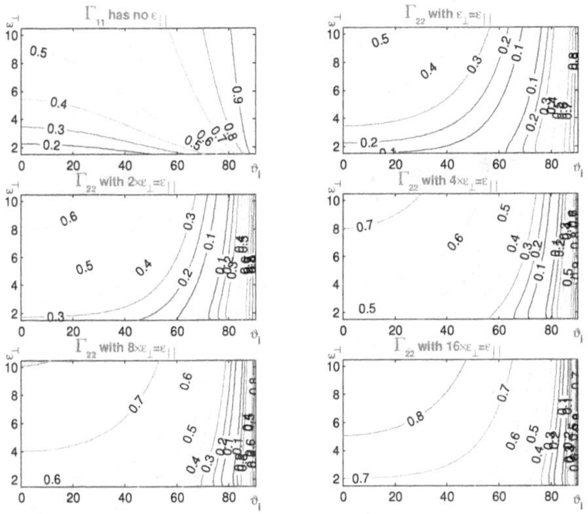

Figure 3.11.: Top View of Figure 3.10.

Figure 3.12 below shows transmission coefficient varying with the angle of incidence and relative permittivity. The higher the values of the incident angle in T_{22} graphs, the higher the transmission coefficient values get. The graphs here fill up the Brewster's angle region in comparison to the earlier gaps in Figure 3.10.

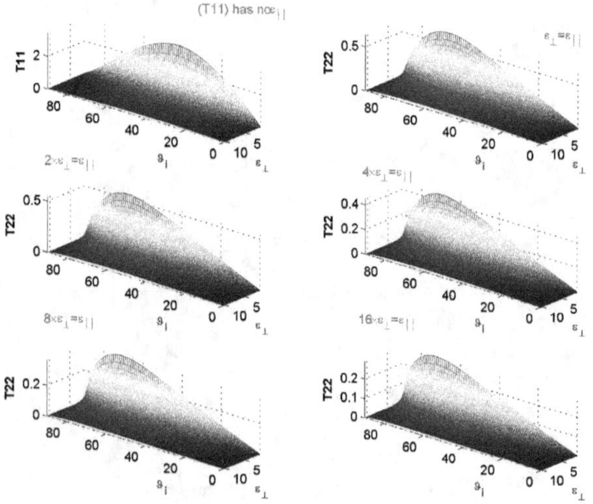

Figure 3.12.: The Transmission Coefficient When the Optic Axis is Parallel to the Plane of Incidence and to the Interface.

And with a top view (see below)

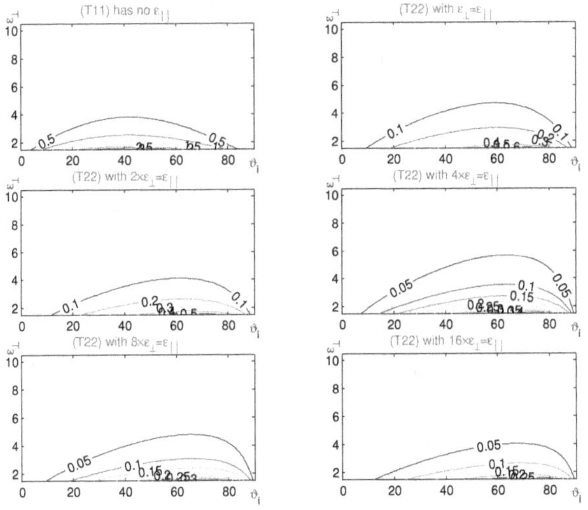

Figure 3.13.: Top View of Figure 3.12.

(2) *Optic Axis Perpendicular to the Plane of Incidence*
In this case, $\hat{\mathbf{c}}$ is parallel to $\mathbf{a}$. Thus $\mathbf{a}\cdot\hat{\mathbf{c}} = |\mathbf{a}|$, $\mathbf{b}\cdot\hat{\mathbf{c}} = \hat{\mathbf{q}}$. $\hat{\mathbf{c}} = 0$, and we have

$$
\begin{aligned}
q_+ &= \sqrt{k_0^2\varepsilon_\perp - \mathbf{a}^2} \qquad k_+^2 = k_0^2\varepsilon_\perp \\
q_- &= \sqrt{k_0^2\varepsilon_\parallel - \mathbf{a}^2} \qquad k_-^2 = k_0^2\varepsilon_\parallel \\
X &= Y = V = L = 0 \\
\Delta_1 &= (q_i + q_-)\mathbf{a}^2\varepsilon_\perp k_0^2(k_0^2\varepsilon_\perp q_i + k_i^2 q_+)
\end{aligned}
\tag{3.235}
$$

Then the transmission and reflection coefficients are reduced to

$$
\begin{aligned}
T_{11} &= 0 \\
T_{12} &= \frac{2k_i q_i|\mathbf{a}|}{(k_0^2\varepsilon_\perp q_i + k_i^2 q_+)} \\
&= \frac{2\cos\theta_i \sin\theta_i}{\varepsilon_\perp \cos\theta_i + \sqrt{\varepsilon_\perp - \sin^2\theta_i}}
\end{aligned}
\tag{3.236}
$$

$$
\begin{aligned}
T_{21} &= \frac{2q_i|\mathbf{a}|}{k_0\varepsilon_\perp(q_i + q_-)} \\
&= \frac{2\cos\theta_i \sin\theta_i}{\varepsilon_\perp \cos\theta_i + \varepsilon_\perp\sqrt{\varepsilon_\parallel - \sin^2\theta_i}} \\
T_{22} &= 0
\end{aligned}
$$

and

$$\begin{aligned}
\Gamma_{11} &= \frac{(q_i - q_-)}{q_i + q_-} \\
&= \frac{\cos\theta_i - \sqrt{\varepsilon_{\|} - \sin^2\theta_i}}{\cos\theta_i + \sqrt{\varepsilon_{\|} - \sin^2\theta_i}} \\
\Gamma_{22} &= \frac{(\varepsilon_{\perp} q_i - \varepsilon_1 q_+)}{(\varepsilon_{\perp} q_i + \varepsilon_1 q_+)} \qquad (3.237) \\
&= \frac{\varepsilon_{\perp}\cos\theta_i - \sqrt{\varepsilon_{\perp} - \sin^2\theta_i}}{\varepsilon_{\perp}\cos\theta_i + \sqrt{\varepsilon_{\perp} - \sin^2\theta_i}}
\end{aligned}$$

$$\Gamma_{12} = \Gamma_{21} = 0 \qquad (3.238)$$

respectively. Here we note

$$k_0^2(\varepsilon_{\perp} q_i \pm \varepsilon_1 q_+) = (q_i \pm q_+)(\mathbf{a}^2 \pm q_i q_+) \qquad (3.239)$$

Figure 3.14 below shows the reflection coefficient varying with the angle of incidence and relative permittivity, where Brewster's angle region is to be noticed only within the Γ_{22} graph.

Γ_{11} values increase with the increase of the incident angle's value; and if $\varepsilon_{\|}$ is to be increased while maintaining $\varepsilon_{\perp}$ fixed, the Γ_{11} values will again increase for all incident angle values. In comparison to Figure 3.16, the transmitted coefficient values would react inversely in such a way that they will decrease with the increase of the $\varepsilon_{\|}$ when $\varepsilon_{\perp}$ is fixed.

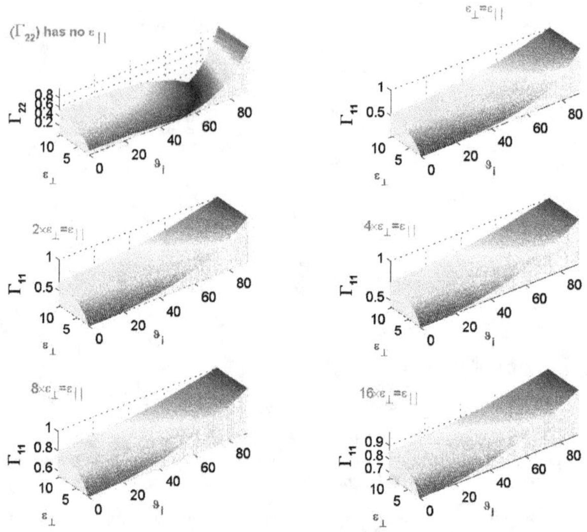

Figure 3.14.: The Reflection Coefficient When the Optic Axis is Perpendicular to the Plane of Incidence and to the Interface.

And with a top view (see below)

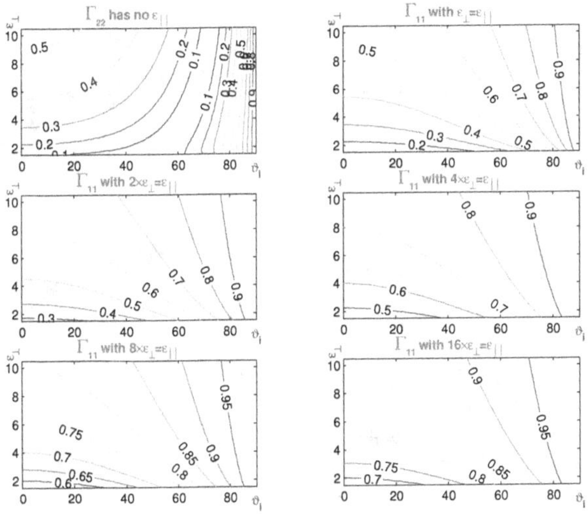

Figure 3.15.: Top view of Figure 3.14.

Figure 3.16 below shows the transmission coefficient varying with the angle of incidence and relative permittivity.

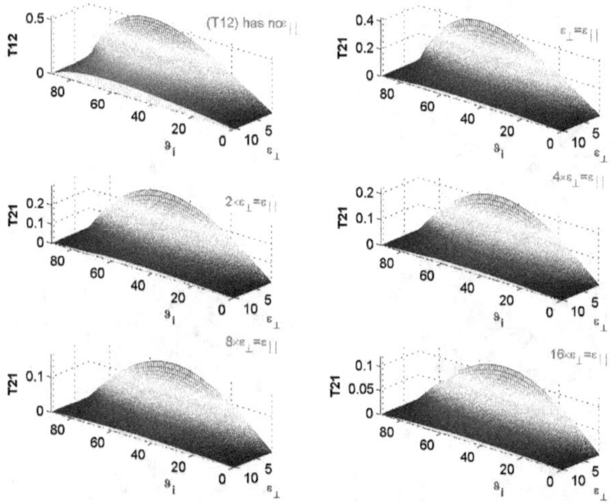

Figure 3.16.: The Transmission Coefficient When the Optic Axis is Perpendicular to the Plane of Incidence and to the Interface.

And with a top view (see below)

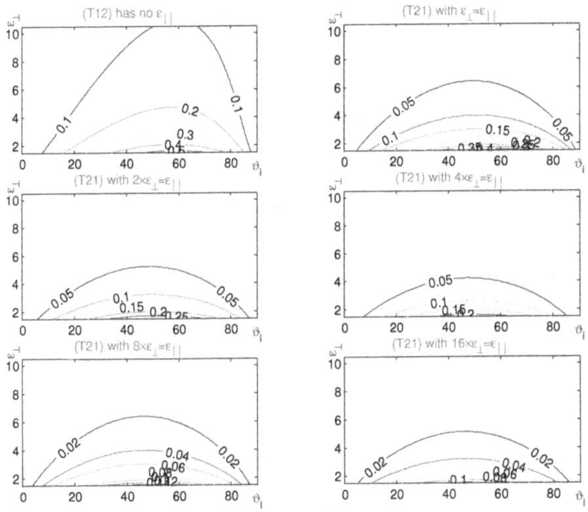

Figure 3.17.: Top view of Figure 3.16.

Optic Axes Parallel to the Plane of Incidence

In this case we have $\mathbf{a} \cdot \widehat{\mathbf{c}}=0$, then we obtain

$$U = Z = V = L = 0 \tag{3.240}$$

$$\Delta_1 = (q_i + q_+)(X + Y) \tag{3.241}$$

Also, from equation (3.186), we note that

$$\mathbf{a} \cdot (\text{adj } \overline{\varepsilon}) \cdot \mathbf{a} = \varepsilon_\perp \varepsilon_\| \mathbf{a}^2 \tag{3.242}$$

Thus

$$\zeta = \{\varepsilon_\perp \varepsilon_\| [k_0^2 (\widehat{\mathbf{q}} \cdot \overline{\varepsilon} \cdot \widehat{\mathbf{q}}) - \mathbf{a}^2]\}^{1/2} \tag{3.243}$$

According to equation (3.187), it follows that

$$\mathbf{k}_- \cdot (\mathbf{a} \times \widehat{\mathbf{c}}) = \frac{1}{\widehat{\mathbf{q}} \cdot \overline{\varepsilon} \cdot \widehat{\mathbf{q}}} [\zeta (\mathbf{b} \cdot \widehat{\mathbf{c}}) - \varepsilon_\| \mathbf{a}^2 (\widehat{\mathbf{q}} \cdot \widehat{\mathbf{c}})] \tag{3.244}$$

Substituting the values of q_+ and q_- in Y yields

$$\begin{aligned} q_+^2 (\mathbf{b} \cdot \widehat{\mathbf{c}}) - q_- \mathbf{a}^2 (\widehat{\mathbf{q}} \cdot \widehat{\mathbf{c}}) &= \frac{1}{\widehat{\mathbf{q}} \cdot \overline{\varepsilon} \cdot \widehat{\mathbf{q}}} \{k_0^2 \varepsilon_\perp (\mathbf{b} \cdot \widehat{\mathbf{c}})(\widehat{\mathbf{q}} \cdot \overline{\varepsilon} \cdot \widehat{\mathbf{q}}) \\ &+ [(\widehat{\mathbf{q}} \cdot \widehat{\mathbf{c}})(\mathbf{b} \cdot \overline{\varepsilon} \cdot \widehat{\mathbf{q}}) \\ &- (\mathbf{b} \cdot \widehat{\mathbf{c}})(\widehat{\mathbf{q}} \cdot \overline{\varepsilon} \cdot \widehat{\mathbf{q}})] \mathbf{a}^2 \\ &- \zeta \mathbf{a}^2 (\widehat{\mathbf{q}} \cdot \widehat{\mathbf{c}})\} \end{aligned} \tag{3.245}$$

But

$$(\widehat{\mathbf{q}} \cdot \widehat{\mathbf{c}})(\mathbf{b} \cdot \overline{\varepsilon} \cdot \widehat{\mathbf{q}}) - (\mathbf{b} \cdot \widehat{\mathbf{c}})(\widehat{\mathbf{q}} \cdot \overline{\varepsilon} \cdot \widehat{\mathbf{q}}) = -\varepsilon_\perp (\mathbf{b} \cdot \widehat{\mathbf{c}})$$

Thus

$$q_+^2 (\mathbf{b} \cdot \widehat{\mathbf{c}}) - q_- \mathbf{a}^2 (\widehat{\mathbf{q}} \cdot \widehat{\mathbf{c}}) = \frac{\zeta}{\varepsilon_\|} (\mathbf{k}_- \cdot \mathbf{a} \times \widehat{\mathbf{c}}) \tag{3.246}$$

where

$$k_0^2 (\widehat{\mathbf{q}} \cdot \overline{\varepsilon} \cdot \widehat{\mathbf{q}}) - \mathbf{a}^2 = \frac{\zeta^2}{\varepsilon_\| \varepsilon_\perp}$$

Consequently, Y becomes

$$Y = \frac{k_i^2 \zeta}{\varepsilon_\parallel}[\mathbf{k}_+ \cdot (\mathbf{a} \times \widehat{\mathbf{c}})][\mathbf{k}_- \cdot (\mathbf{a} \times \widehat{\mathbf{c}})]$$

giving

$$\begin{aligned} T_{11} &= -\frac{2q_i \mathbf{a}^2}{(q_i + q_+)[\mathbf{k}_+ \cdot (\mathbf{a} \times \widehat{\mathbf{c}})]} \\ T_{22} &= \frac{2k_0 q_i \mathbf{a}^2 k_i \varepsilon_\parallel}{[\mathbf{k}_- \cdot (\mathbf{a} \times \widehat{\mathbf{c}})](k_0^2 \varepsilon_\perp \varepsilon_\parallel q_i + k_i^2 \zeta)} \\ T_{12} &= T_{21} = 0 \end{aligned} \tag{3.247}$$

and the reflection coefficients

$$\begin{aligned} \Gamma_{11} &= \frac{q_i - q_+}{q_i + q_+} \\ \Gamma_{22} &= \frac{\sqrt{\varepsilon_\perp \varepsilon_\parallel}\ q_i - \varepsilon_1 \sqrt{k_0^2(\widehat{\mathbf{q}} \cdot \overline{\varepsilon} \cdot \widehat{\mathbf{q}}) - \mathbf{a}^2}}{\sqrt{\varepsilon_\perp \varepsilon_\parallel}\ q_i + \varepsilon_1 \sqrt{k_0^2(\widehat{\mathbf{q}} \cdot \overline{\varepsilon} \cdot \widehat{\mathbf{q}}) - \mathbf{a}^2}} \\ \Gamma_{21} &= \Gamma_{12} = 0 \end{aligned} \tag{3.248}$$

When there is a certain angle β between the optic axis and the normal to the interface:

This is the case when $\alpha = 90^o - \beta$, $\mathbf{b} \cdot \widehat{\mathbf{c}} = |a| \cos\alpha$ and $\widehat{\mathbf{q}} \cdot \widehat{\mathbf{c}} = \cos\beta$

The transmission coefficient becomes

$$\begin{aligned} T_{11} &= \frac{-2\cos\theta_i \sin^2\theta_i}{(\cos\theta_i + \sqrt{\varepsilon_\perp - \sin^2\theta_i})} \\ &\quad * \frac{1}{[-\sin^2\theta_i \cos\beta + \sin\theta_i \sqrt{\varepsilon_\perp - \sin^2\theta_i}\ \cos\alpha]} \\ T_{22} &= \frac{2\cos\theta_i \varepsilon_\parallel k_0^2 \sin^2\theta_i}{[q_- |a| \cos\alpha - |a|^2 \cos\beta](\varepsilon_\perp \varepsilon_\parallel k_0 \cos\theta_i + \zeta)} \\ T_{12} &= T_{21} = 0 \end{aligned} \tag{3.249}$$

where

$$q_- = \frac{-(\varepsilon_{\|} - \varepsilon_{\perp})|a| \cos\beta \cos\alpha}{\varepsilon_{\perp} + (\varepsilon_{\|} - \varepsilon_{\perp}) \cos^2\beta} + \frac{\sqrt{\varepsilon_{\perp}\varepsilon_{\|}[k_0^2(\varepsilon_{\perp} + (\varepsilon_{\|} - \varepsilon_{\perp})\cos^2\beta) - k_0^2 \sin^2\theta_i]}}{\varepsilon_{\perp} + (\varepsilon_{\|} - \varepsilon_{\perp}) \cos^2\beta}$$

and the reflection coefficient would be

$$\Gamma_{11} = \frac{\cos\theta_i - \sqrt{\varepsilon_{\perp} - \sin^2\theta_i}}{\cos\theta_i + \sqrt{\varepsilon_{\perp} - \sin^2\theta_i}}$$

$$\Gamma_{22} = \frac{\sqrt{\varepsilon_{\perp}\varepsilon_{\|}}\ \cos_i - \sqrt{\varepsilon_{\perp} + (\varepsilon_{\|} - \varepsilon_{\perp})\cos^2\beta - \sin^2\theta_i}}{\sqrt{\varepsilon_{\perp}\varepsilon_{\|}}\ \cos_i + \sqrt{\varepsilon_{\perp} + (\varepsilon_{\|} - \varepsilon_{\perp})\cos^2\beta - \sin^2\theta_i}} \quad (3.250)$$

$$\Gamma_{21} = \Gamma_{12} = 0$$

Solving for $\beta = 30^o$, we obtain Figure 3.18 below which shows the reflection coefficient varying with the angle of incidence and relative permittivity.

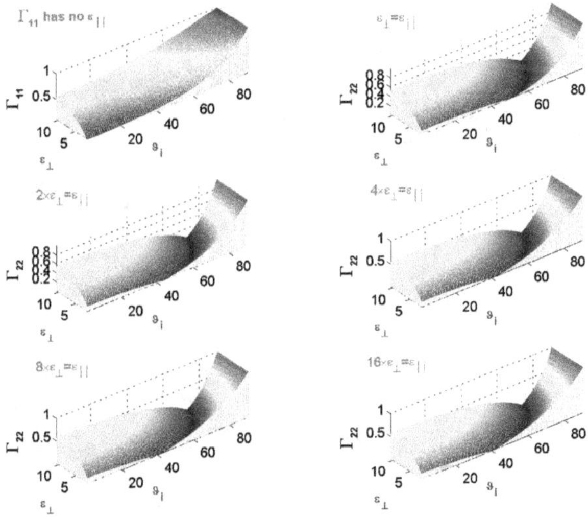

Figure 3.18.: The Reflection Coefficient When the Optic Axis is Parallel to the Plane of Incidence and $\beta = 30^o$.

And with a top view (see below)

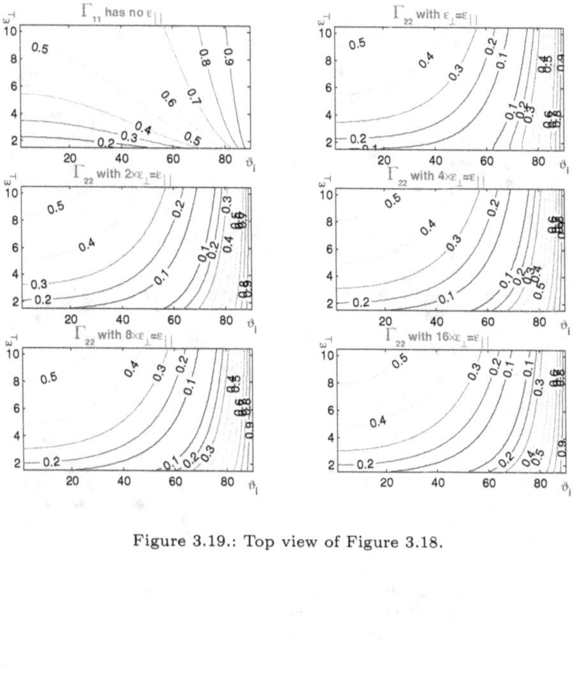

Figure 3.19.: Top view of Figure 3.18.

Figure 3.20 below shows the transmission coefficient varying with the angle of incidence and relative permittivity. It fills up the empty region of the reflection coefficient graphs in Figure 3.18.

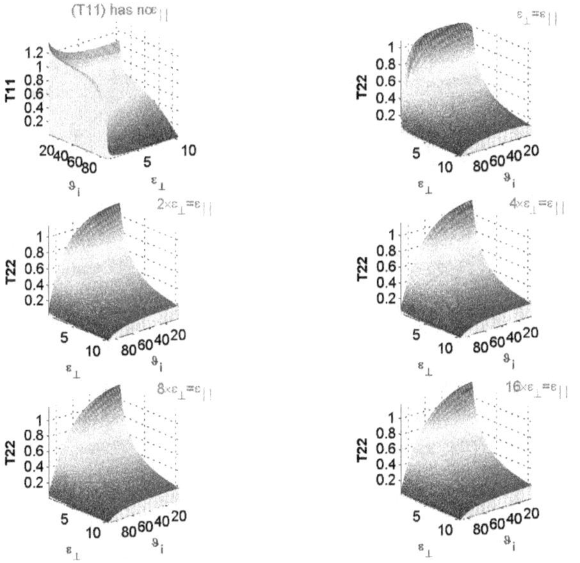

Figure 3.20.: The Transmission Coefficient When the Optic Axis is Parallel to the Plane of Incidence and $\beta = 30^o$.

And with a top view (see below)

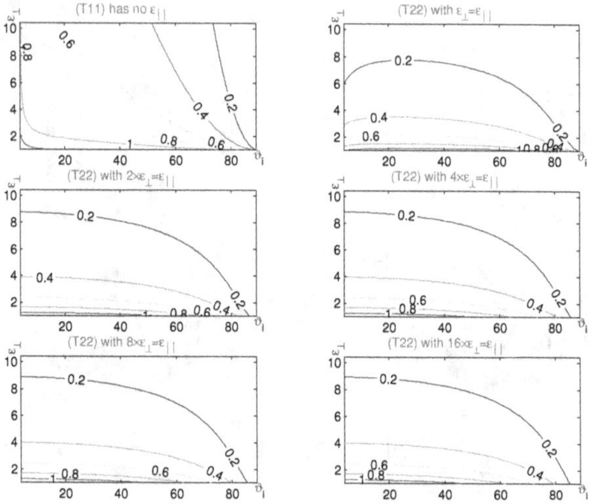

Figure 3.21.: Top view of Figure 3.20.

Solving for $\beta = 60^o$, we obtain Figure 3.22 below which shows the reflection coefficient varying with the angle of incidence and relative permittivity.

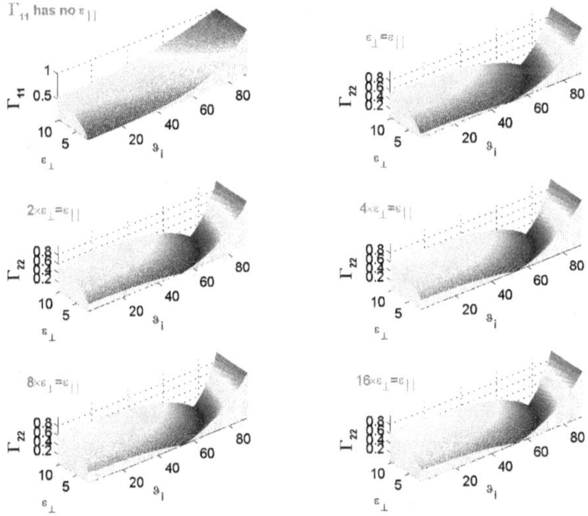

Figure 3.22.: Reflection coefficient when the optic axis is parallel to the plane of incidence and $\beta = 60^o$.

And with a top view (see below)

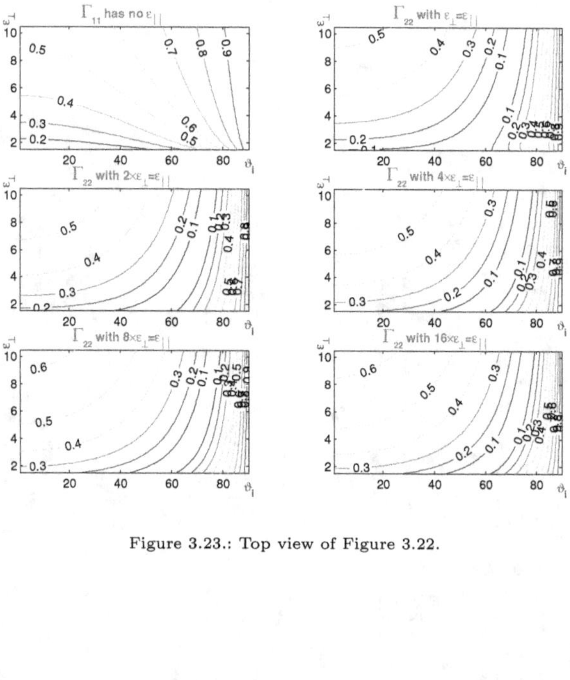

Figure 3.23.: Top view of Figure 3.22.

Figure 3.24 below shows the transmission coefficient varying with the angle of incidence and relative permittivity.

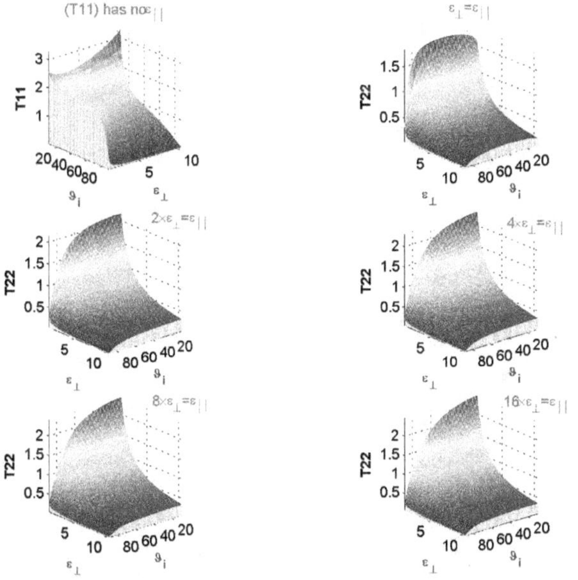

Figure 3.24.: The Transmission Coefficient When the Optic Axis is Parallel to the Plane of Incidence and $\beta = 60^o$.

And with a top view (see below)

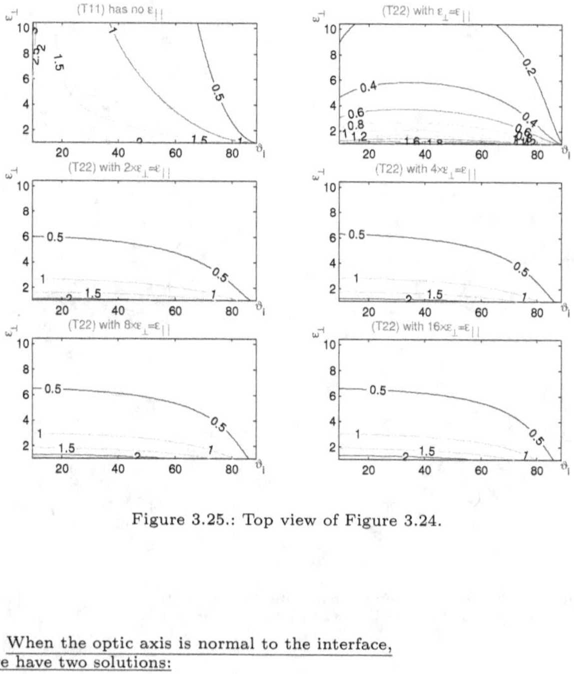

Figure 3.25.: Top view of Figure 3.24.

When the optic axis is normal to the interface, we have two solutions:

(A) *Oblique incidence*

In this case $\widehat{\mathbf{c}} = \widehat{\mathbf{q}}$, $\mathbf{a} \cdot \widehat{\mathbf{c}} = \mathbf{b} \cdot \widehat{\mathbf{c}} = 0$, $\mathbf{a} \times \widehat{\mathbf{c}} = -\mathbf{b}$, and

$$\begin{aligned}
\mathbf{k}_+ \cdot (\mathbf{a} \times \widehat{\mathbf{c}}) &= \mathbf{k}_- \cdot (\mathbf{a} \times \widehat{\mathbf{c}}) = -\mathbf{a}^2 \\
(\widehat{\mathbf{q}} \cdot \overline{\varepsilon} \cdot \widehat{\mathbf{q}}) &= \varepsilon_{\|} \\
(\mathbf{b} \cdot \overline{\varepsilon} \cdot \widehat{\mathbf{q}}) &= 0
\end{aligned} \tag{3.251}$$

giving

$$\begin{aligned}
\zeta = \sqrt{\varepsilon_\perp \varepsilon_\|}\sqrt{k_0^2 \varepsilon_\| - \mathbf{a}^2} \\
q_- = \frac{\zeta}{\varepsilon_\|} \\
X \pm Y = k_0^2 \mathbf{a}^4 (\varepsilon_\perp q_i \pm \varepsilon_1 q_-)
\end{aligned} \tag{3.252}$$

with the transmission coefficients

$$\begin{aligned}
T_{11} &= \frac{2q_i}{q_i + q_+} \\
&= \frac{2\cos\theta_i}{\cos\theta_i + \sqrt{\varepsilon_\perp - \sin^2\theta_i}} \\
T_{22} &= -\frac{2k_0 k_i q_i}{k_0^2 \varepsilon_\perp q_i + k_i^2 q_-} \\
&= -\frac{2\cos\theta_i}{\varepsilon_\perp \cos\theta_i + \sqrt{\frac{\varepsilon_\perp}{\varepsilon_\|}}\sqrt{\varepsilon_\| - \sin^2\theta_i}} \\
T_{12} &= T_{21} = 0
\end{aligned} \tag{3.253}$$

and the reflection coefficients

$$\begin{aligned}
\Gamma_{11} &= \frac{q_i - q_+}{q_i + q_+} \\
&= \frac{\cos\theta_i - \sqrt{\varepsilon_\perp - \sin^2\theta_i}}{\cos\theta_i + \sqrt{\varepsilon_\perp - \sin^2\theta_i}} \\
\Gamma_{22} &= \frac{\sqrt{\varepsilon_\perp \varepsilon_\parallel}\; q_i - \varepsilon_1 \sqrt{k_0^2 \varepsilon_\parallel - \mathbf{a}^2}}{\sqrt{\varepsilon_\perp \varepsilon_\parallel}\; q_i + \varepsilon_1 \sqrt{k_0^2 \varepsilon_\parallel - \mathbf{a}^2}} \\
&= \frac{\sqrt{\varepsilon_\perp \varepsilon_\parallel}\; \cos\theta_i - \sqrt{\varepsilon_\parallel - \sin^2\theta_i}}{\sqrt{\varepsilon_\perp \varepsilon_\parallel}\; \cos\theta_i + \sqrt{\varepsilon_\parallel - \sin^2\theta_i}} \\
\Gamma_{21} &= \Gamma_{12} = 0
\end{aligned} \tag{3.254}$$

Figure 3.26 below shows the reflection coefficient varying with the angle of incidence and relative permittivity.

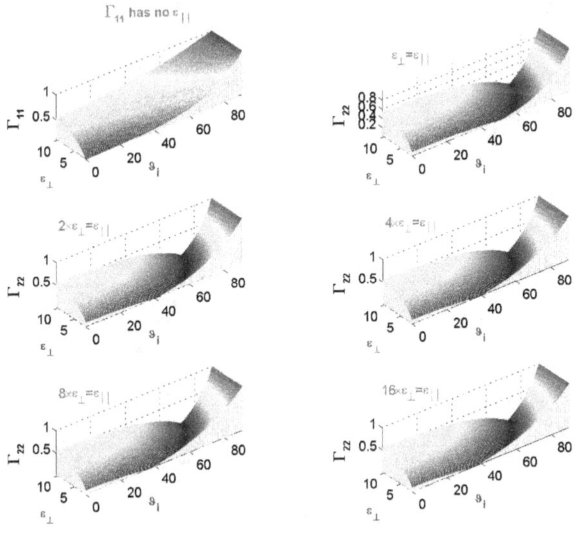

Figure 3.26.: The Reflection Coefficient When the Optic Axis is Parallel to the Plane of Incidence and $\widehat{\mathbf{c}} = \widehat{\mathbf{q}}$.

And with a top view (see below)

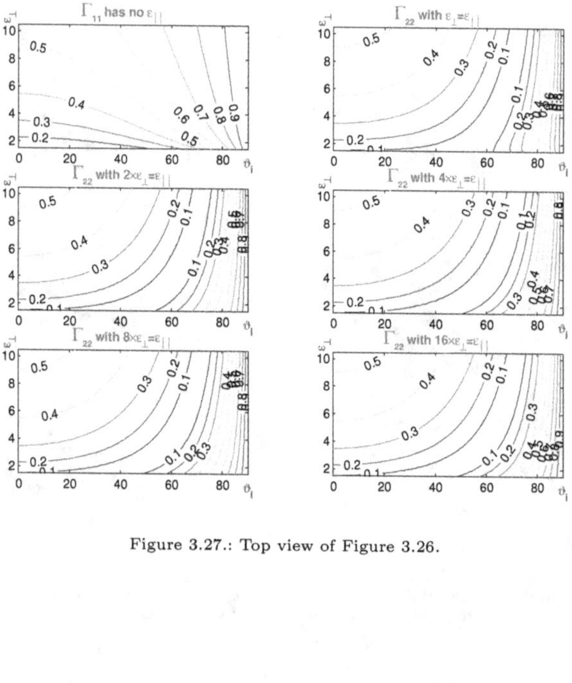

Figure 3.27.: Top view of Figure 3.26.

Figure 3.28 below shows the transmission coefficient varying with the angle of incidence and relative permittivity.

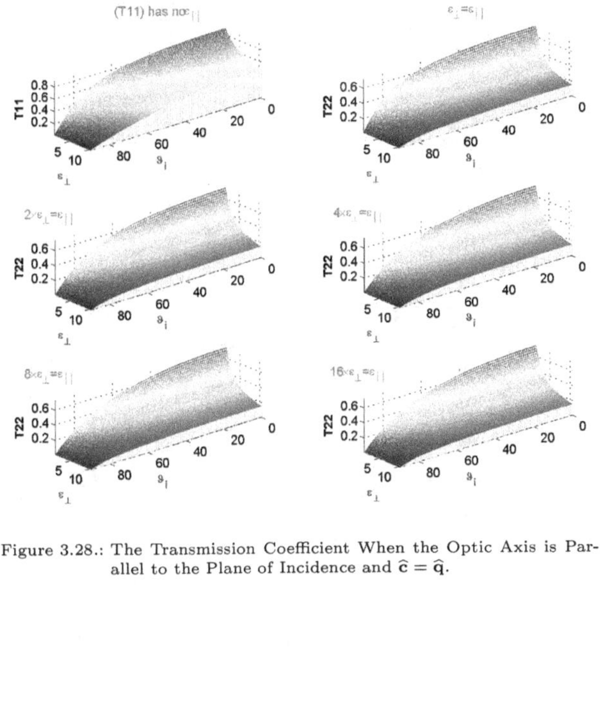

Figure 3.28.: The Transmission Coefficient When the Optic Axis is Parallel to the Plane of Incidence and $\widehat{\mathbf{c}} = \widehat{\mathbf{q}}$.

And with a top view (see below)

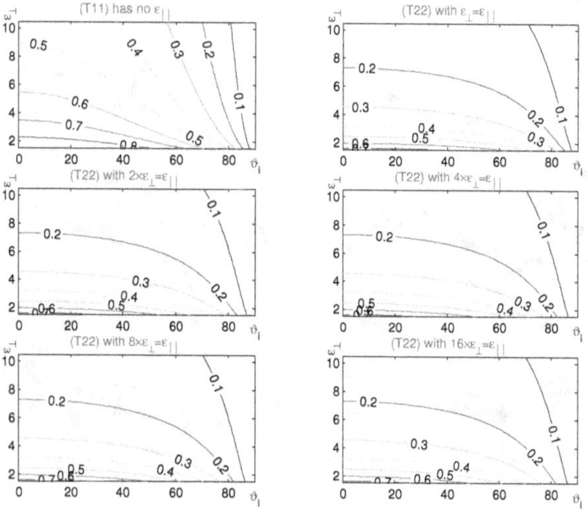

Figure 3.29.: Top view of Figure 3.28.

(B) *Normal incidence*

For $\widehat{\mathbf{c}} = \widehat{\mathbf{q}}$ we see that formulas (3.210) to (3.212) are no longer useful. In this case we have $\mathbf{k}_+ = \mathbf{k}_- = k_0\sqrt{\varepsilon_\perp}\;\widehat{\mathbf{c}} = \mathbf{k}_t$. Hence $\mathbf{k}_t \cdot \mathbf{D}_{0t} = 0$ implies $\widehat{\mathbf{c}} \cdot \mathbf{E}_{0t} = 0$. Following the derivation of normal incidence on the interface of two isotropic media, we obtain

$$\mathbf{E}_{0r} = \frac{k_i - k_t}{k_i + k_t}\mathbf{E}_{0i} \qquad \mathbf{H}_{0r} = \frac{-k_i + k_t}{k_i + k_t}\mathbf{H}_{0i}$$

$$\mathbf{E}_{0t} = \frac{2k_i}{k_i + k_t}\mathbf{E}_{0i} \qquad \mathbf{H}_{0t} = \frac{2k_t}{k_i + k_t}\mathbf{H}_{0i} \tag{3.255}$$

3.8. Reflection and Transmission of Plane Waves Propagating in Lossless, Nonmagnetic and Unbounded Isotropic Media through a Plate of Lossless, Nonmagnetic and Bounded Homogeneous Uniaxial Media

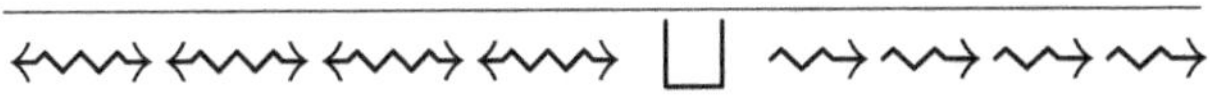

3.8.1. Optic Axis Perpendicular to the Normal of the Interface

For a uniform plane wave propagating through an anisotropic uniaxial plate perpendicular to its interface and to the optic axis as shown in Figure 3.30 below.

From equations (3.210) to (3.212) we have the following field vectors

$$\mathbf{E}_i = A_\perp(\widehat{\mathbf{q}} \times \widehat{\mathbf{c}})e^{-jk_0 z} + A_\parallel[\widehat{\mathbf{q}} \times (\widehat{\mathbf{q}} \times \widehat{\mathbf{c}})]e^{-jk_0 z} \tag{3.256}$$

$$\mathbf{E}_r = B_\perp(\widehat{\mathbf{q}} \times \widehat{\mathbf{c}})e^{jk_0 z} + B_\parallel[\widehat{\mathbf{q}} \times (\widehat{\mathbf{q}} \times \widehat{\mathbf{c}})]e^{jk_0 z} \tag{3.257}$$

$$\mathbf{E}_+ = C_+(\widehat{\mathbf{q}} \times \widehat{\mathbf{c}})e^{-jk_+ z} \tag{3.258}$$

$$\mathbf{E}_{+r} = D_+(\widehat{\mathbf{q}} \times \widehat{\mathbf{c}})e^{jk_+ z} \tag{3.259}$$

$$\mathbf{E}_- = \varepsilon_\perp C_- \widehat{\mathbf{c}} e^{-jk_- z} \tag{3.260}$$

$$\mathbf{E}_{-r} = \varepsilon_\perp D_- \widehat{\mathbf{c}} e^{jk_- z} \tag{3.261}$$

$$\mathbf{E}_T = F_\perp(\widehat{\mathbf{q}} \times \widehat{\mathbf{c}})e^{-jk_0 z} + F_\parallel[\widehat{\mathbf{q}} \times (\widehat{\mathbf{q}} \times \widehat{\mathbf{c}})]e^{-jk_0 z} \tag{3.262}$$

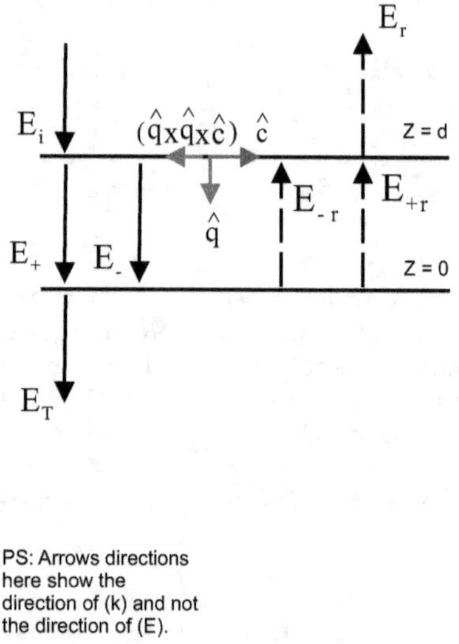

Figure 3.30.: Uniform Plane Wave Propagating through an Uniaxial Plate that is Perpendicular to its Interface and Optic Axis.

and

$$\mathbf{H}_i = \frac{1}{\omega\mu_0}(\mathbf{k}_i \times \mathbf{E}_i) = \sqrt{\frac{\varepsilon_0\varepsilon_1}{\mu_0}}\{A_\perp[\widehat{\mathbf{q}} \times (\widehat{\mathbf{q}} \times \widehat{\mathbf{c}})]e^{-jk_0z} - A_\parallel(\widehat{\mathbf{q}} \times \widehat{\mathbf{c}})e^{-jk_0z}\} \tag{3.263}$$

$$\mathbf{H}_r = \frac{1}{\omega\mu_0}(\mathbf{k}_r \times \mathbf{E}_r) = \sqrt{\frac{\varepsilon_0\varepsilon_1}{\mu_0}}\{-B_\perp[\widehat{\mathbf{q}} \times (\widehat{\mathbf{q}} \times \widehat{\mathbf{c}})]e^{jk_0z} + B_\parallel(\widehat{\mathbf{q}} \times \widehat{\mathbf{c}})e^{jk_0z}\} \tag{3.264}$$

$$\begin{aligned}\mathbf{H}_+ &= \frac{1}{\omega\mu_0}(\mathbf{k}_+ \times \mathbf{E}_+) \\ &= \sqrt{\frac{\varepsilon_0\varepsilon_\perp}{\mu_0}} C_+[\widehat{\mathbf{q}} \times (\widehat{\mathbf{q}} \times \widehat{\mathbf{c}})]e^{-jk_+z} \end{aligned} \tag{3.265}$$

$$\begin{aligned}\mathbf{H}_{+r} &= \frac{1}{\omega\mu_0}(\mathbf{k}_{+r} \times \mathbf{E}_{+r}) \\ &= -\sqrt{\frac{\varepsilon_0\varepsilon_\perp}{\mu_0}} D_+[\widehat{\mathbf{q}} \times (\widehat{\mathbf{q}} \times \widehat{\mathbf{c}})]e^{jk_+z} \end{aligned} \tag{3.266}$$

$$\begin{aligned}\mathbf{H}_- &= \frac{1}{\omega\mu_0}(\mathbf{k}_- \times \mathbf{E}_-) \\ &= \varepsilon_\perp\sqrt{\frac{\varepsilon_0\varepsilon_\parallel}{\mu_0}} C_-(\widehat{\mathbf{q}} \times \widehat{\mathbf{c}})e^{-jk_-z} \end{aligned} \tag{3.267}$$

$$\begin{aligned}\mathbf{H}_{-r} &= \frac{1}{\omega\mu_0}(\mathbf{k}_- \times \mathbf{E}_-) \\ &= -\varepsilon_\perp\sqrt{\frac{\varepsilon_0\varepsilon_\parallel}{\mu_0}} D_-(\widehat{\mathbf{q}} \times \widehat{\mathbf{c}})e^{jk_-z} \end{aligned} \tag{3.268}$$

$$\begin{aligned}\mathbf{H}_T &= \frac{1}{\omega\mu_0}(\mathbf{k}_T \times \mathbf{E}_T) \\ &= \sqrt{\frac{\varepsilon_0\varepsilon_1}{\mu_0}}\{F_\perp[\widehat{\mathbf{q}} \times (\widehat{\mathbf{q}} \times \widehat{\mathbf{c}})]e^{-jk_0z} - F_\parallel(\widehat{\mathbf{q}} \times \widehat{\mathbf{c}})e^{-jk_0z}\} \end{aligned} \tag{3.269}$$

with the wave vectors

$$\begin{gathered}\mathbf{k}_i = k_i\widehat{\mathbf{q}} = -\mathbf{k}_r, \quad \mathbf{k}_+ = k_+\widehat{\mathbf{q}}, \quad \mathbf{k}_- = k_-\widehat{\mathbf{q}} \\ \mathbf{k}_{+r} = -k_+\widehat{\mathbf{q}}, \quad \mathbf{k}_{-r} = -k_-\widehat{\mathbf{q}} \end{gathered} \tag{3.270}$$

where

$$\begin{gathered} k_i = k_0\sqrt{\varepsilon_1}, \quad k_+ = k_0\sqrt{\varepsilon_\perp}, \\ k_- = k_0\sqrt{\frac{\varepsilon_\perp\varepsilon_\parallel}{\widehat{\mathbf{q}}\cdot\overline{\varepsilon}\cdot\widehat{\mathbf{q}}}} = k_0\sqrt{\varepsilon_\parallel} \end{gathered} \tag{3.271}$$

Substituting the electric field vectors in the boundary conditions when z=d (first interface), we obtain

$$\begin{aligned} A_\perp & \quad (\hat{\mathbf{q}} \times \hat{\mathbf{c}}) e^{-jk_0 d} + A_\| [\hat{\mathbf{q}} \times (\hat{\mathbf{q}} \times \hat{\mathbf{c}})] e^{-jk_0 d} \\ + & \quad B_\perp (\hat{\mathbf{q}} \times \hat{\mathbf{c}}) e^{jk_0 d} + B_\| [\hat{\mathbf{q}} \times (\hat{\mathbf{q}} \times \hat{\mathbf{c}})] e^{jk_0 d} \\ - & \quad C_+ (\hat{\mathbf{q}} \times \hat{\mathbf{c}}) e^{-jk_+ d} - D_+ (\hat{\mathbf{q}} \times \hat{\mathbf{c}}) e^{jk_+ d} \\ - & \quad \varepsilon_\perp C_- \hat{\mathbf{c}} e^{-jk_- d} - \varepsilon_\perp D_- \hat{\mathbf{c}} e^{jk_- d} = \alpha \hat{\mathbf{q}} \end{aligned}$$

we cross-premultiply the former equation with $\hat{\mathbf{q}}$ to get

$$\begin{aligned} A_\perp & \quad (\hat{\mathbf{q}} \times \hat{\mathbf{q}} \times \hat{\mathbf{c}}) e^{-jk_0 d} - A_\| (\hat{\mathbf{q}} \times \hat{\mathbf{c}}) e^{-jk_0 d} \\ + & \quad B_\perp (\hat{\mathbf{q}} \times \hat{\mathbf{q}} \times \hat{\mathbf{c}}) e^{jk_0 d} - B_\| (\hat{\mathbf{q}} \times \hat{\mathbf{c}}) e^{jk_0 d} \\ - & \quad C_+ (\hat{\mathbf{q}} \times \hat{\mathbf{q}} \times \hat{\mathbf{c}}) e^{-jk_+ d} - D_+ (\hat{\mathbf{q}} \times \hat{\mathbf{q}} \times \hat{\mathbf{c}}) e^{jk_+ d} \\ - & \quad \varepsilon_\perp C_- (\hat{\mathbf{q}} \times \hat{\mathbf{c}}) e^{-jk_- d} - \varepsilon_\perp D_- (\hat{\mathbf{q}} \times \hat{\mathbf{c}}) e^{jk_- d} \\ = & \quad \mathbf{0} \end{aligned} \tag{3.272}$$

and from equation (2.40),

$$\begin{aligned} \sqrt{\frac{\varepsilon_0 \varepsilon_1}{\mu_0}} & \quad \{A_\perp [\hat{\mathbf{q}} \times (\hat{\mathbf{q}} \times \hat{\mathbf{c}})] e^{-jk_0 d} - A_\| (\hat{\mathbf{q}} \times \hat{\mathbf{c}}) e^{-jk_0 d}\} \\ + & \quad \sqrt{\frac{\varepsilon_0 \varepsilon_1}{\mu_0}} \{-B_\perp [\hat{\mathbf{q}} \times (\hat{\mathbf{q}} \times \hat{\mathbf{c}})] e^{jk_0 d} + B_\| (\hat{\mathbf{q}} \times \hat{\mathbf{c}}) e^{jk_0 d}\} \\ - & \quad \sqrt{\frac{\varepsilon_0 \varepsilon_\perp}{\mu_0}} C_+ [\hat{\mathbf{q}} \times (\hat{\mathbf{q}} \times \hat{\mathbf{c}})] e^{-jk_+ d} \\ + & \quad \sqrt{\frac{\varepsilon_0 \varepsilon_\perp}{\mu_0}} D_+ [\hat{\mathbf{q}} \times (\hat{\mathbf{q}} \times \hat{\mathbf{c}})] e^{jk_+ d} \\ - & \quad \varepsilon_\perp \sqrt{\frac{\varepsilon_0 \varepsilon_\|}{\mu_0}} C_- (\hat{\mathbf{q}} \times \hat{\mathbf{c}}) e^{-jk_- d} \\ + & \quad \varepsilon_\perp \sqrt{\frac{\varepsilon_0 \varepsilon_\|}{\mu_0}} D_- (\hat{\mathbf{q}} \times \hat{\mathbf{c}}) e^{jk_- d} = \mathbf{0} \end{aligned}$$

which can be simplified to

$$\begin{aligned}
& A_\perp[\widehat{\mathbf{q}} \times (\widehat{\mathbf{q}} \times \widehat{\mathbf{c}})]e^{-jk_0 d} - A_\|(\widehat{\mathbf{q}} \times \widehat{\mathbf{c}})e^{-jk_0 d} \\
& -B_\perp[\widehat{\mathbf{q}} \times (\widehat{\mathbf{q}} \times \widehat{\mathbf{c}})]e^{jk_0 d} + B_\|(\widehat{\mathbf{q}} \times \widehat{\mathbf{c}})e^{jk_0 d} \\
- & \sqrt{\frac{\varepsilon_\perp}{\varepsilon_1}} C_+[\widehat{\mathbf{q}} \times (\widehat{\mathbf{q}} \times \widehat{\mathbf{c}})]e^{-jk_+ d} \\
+ & \sqrt{\frac{\varepsilon_\perp}{\varepsilon_1}} D_+[\widehat{\mathbf{q}} \times (\widehat{\mathbf{q}} \times \widehat{\mathbf{c}})]e^{jk_+ d} \\
- & \varepsilon_\perp \sqrt{\frac{\varepsilon_\|}{\varepsilon_1}} C_-(\widehat{\mathbf{q}} \times \widehat{\mathbf{c}})e^{-jk_- d} \\
+ & \varepsilon_\perp \sqrt{\frac{\varepsilon_\|}{\varepsilon_1}} D_-(\widehat{\mathbf{q}} \times \widehat{\mathbf{c}})e^{jk_- d} = \mathbf{0}
\end{aligned} \tag{3.273}$$

Then substituting the field vectors in the boundary conditions when z=0 (second interface), we obtain

$$\begin{aligned}
& C_+(\widehat{\mathbf{q}} \times \widehat{\mathbf{c}}) + D_+(\widehat{\mathbf{q}} \times \widehat{\mathbf{c}}) + \varepsilon_\perp C_- \widehat{\mathbf{c}} \\
+ & \varepsilon_\perp D_- \widehat{\mathbf{c}} - F_\perp(\widehat{\mathbf{q}} \times \widehat{\mathbf{c}}) - F_\|[\widehat{\mathbf{q}} \times (\widehat{\mathbf{q}} \times \widehat{\mathbf{c}})] = \beta \widehat{\mathbf{q}}
\end{aligned}$$

We cross-premultiply the former equation with $\widehat{\mathbf{q}}$ to get

$$\begin{aligned}
& C_+(\widehat{\mathbf{q}} \times \widehat{\mathbf{q}} \times \widehat{\mathbf{c}}) + D_+(\widehat{\mathbf{q}} \times \widehat{\mathbf{q}} \times \widehat{\mathbf{c}}) + \varepsilon_\perp C_-(\widehat{\mathbf{q}} \times \widehat{\mathbf{c}}) \\
+ & \varepsilon_\perp D_-(\widehat{\mathbf{q}} \times \widehat{\mathbf{c}}) - F_\perp(\widehat{\mathbf{q}} \times \widehat{\mathbf{q}} \times \widehat{\mathbf{c}}) \\
+ & F_\|(\widehat{\mathbf{q}} \times \widehat{\mathbf{c}}) = \mathbf{0}
\end{aligned} \tag{3.274}$$

and from equation (2.40),

$$\begin{aligned}
& \sqrt{\frac{\varepsilon_0 \varepsilon_\perp}{\mu_0}} C_+[\widehat{\mathbf{q}} \times (\widehat{\mathbf{q}} \times \widehat{\mathbf{c}})] - \sqrt{\frac{\varepsilon_0 \varepsilon_\perp}{\mu_0}} D_+[\widehat{\mathbf{q}} \times (\widehat{\mathbf{q}} \times \widehat{\mathbf{c}})] \\
+ & \sqrt{\frac{\varepsilon_0 \varepsilon_\|}{\mu_0}} C_-(\widehat{\mathbf{q}} \times \widehat{\mathbf{c}})\varepsilon_\perp - \varepsilon_\perp \sqrt{\frac{\varepsilon_0 \varepsilon_\|}{\mu_0}} D_-(\widehat{\mathbf{q}} \times \widehat{\mathbf{c}}) \\
- & \sqrt{\frac{\varepsilon_0 \varepsilon_1}{\mu_0}} \{F_\perp[\widehat{\mathbf{q}} \times (\widehat{\mathbf{q}} \times \widehat{\mathbf{c}})] - F_\|(\widehat{\mathbf{q}} \times \widehat{\mathbf{c}})\} = \mathbf{0}
\end{aligned}$$

Which can be simplified into

$$\begin{aligned}
& C_+[\widehat{\mathbf{q}} \times (\widehat{\mathbf{q}} \times \widehat{\mathbf{c}})] - D_+[\widehat{\mathbf{q}} \times (\widehat{\mathbf{q}} \times \widehat{\mathbf{c}})] \\
+ & \sqrt{\frac{\varepsilon_{\|}}{\varepsilon_{\perp}}} C_-(\widehat{\mathbf{q}} \times \widehat{\mathbf{c}})\varepsilon_{\perp} - \varepsilon_{\perp}\sqrt{\frac{\varepsilon_{\|}}{\varepsilon_{\perp}}} D_-(\widehat{\mathbf{q}} \times \widehat{\mathbf{c}}) \\
- & \sqrt{\frac{\varepsilon_1}{\varepsilon_{\perp}}} F_{\perp}[\widehat{\mathbf{q}} \times (\widehat{\mathbf{q}} \times \widehat{\mathbf{c}})] + \sqrt{\frac{\varepsilon_1}{\varepsilon_{\perp}}} F_{\|}(\widehat{\mathbf{q}} \times \widehat{\mathbf{c}}) \\
= & \mathbf{0}
\end{aligned} \tag{3.275}$$

Since $\widehat{\mathbf{q}} \times \widehat{\mathbf{c}}$ and $\widehat{\mathbf{q}} \times (\widehat{\mathbf{q}} \times \widehat{\mathbf{c}})$ are two linearly independent vectors[17], we can separate the solutions so that

$$\begin{aligned}
A_{\perp} e^{-jk_0 d} & + B_{\perp} e^{jk_0 d} \\
& - C_+ e^{-jk_+ d} - D_+ e^{jk_+ d} = 0
\end{aligned} \tag{3.276}$$

$$\begin{aligned}
-A_{\|} e^{-jk_0 d} & - B_{\|} e^{jk_0 d} \\
-\varepsilon_{\perp} C_- e^{-jk_- d} & - \varepsilon_{\perp} D_- e^{jk_- d} = 0
\end{aligned} \tag{3.277}$$

$$\begin{aligned}
A_{\perp} e^{-jk_0 d} & - B_{\perp} e^{jk_0 d} \\
-\sqrt{\frac{\varepsilon_{\perp}}{\varepsilon_1}} C_+ e^{-jk_+ d} & + \sqrt{\frac{\varepsilon_{\perp}}{\varepsilon_1}} D_+ e^{jk_+ d} = 0
\end{aligned} \tag{3.278}$$

$$\begin{aligned}
-A_{\|} e^{-jk_0 d} & + B_{\|} e^{jk_0 d} \\
-\varepsilon_{\perp}\sqrt{\frac{\varepsilon_{\|}}{\varepsilon_1}} C_- e^{-jk_- d} & + \varepsilon_{\perp}\sqrt{\frac{\varepsilon_{\|}}{\varepsilon_1}} D_- e^{jk_- d} = 0
\end{aligned} \tag{3.279}$$

$$C_+ + D_+ - F_{\perp} = 0 \tag{3.280}$$

$$\varepsilon_{\perp} C_- + F_{\|} + \varepsilon_{\perp} D_- = 0 \tag{3.281}$$

$$C_+ - D_+ - \sqrt{\frac{\varepsilon_1}{\varepsilon_{\perp}}} F_{\perp} = 0 \tag{3.282}$$

[17] See Appendix G1.

$$\varepsilon_{\perp}\sqrt{\frac{\varepsilon_{\|}}{\varepsilon_{\perp}}}C_{-} - \varepsilon_{\perp}\sqrt{\frac{\varepsilon_{\|}}{\varepsilon_{\perp}}}D_{-} + \sqrt{\frac{\varepsilon_{1}}{\varepsilon_{\perp}}}F_{\|} = 0 \tag{3.283}$$

We can find some ratios out of this equation system which has 8 unknowns (where $A_{\perp}$ and $A_{\|}$ are given.) and 8 equations.
Summing up equation (3.282) with equation (3.284) gives

$$\begin{aligned} -\varepsilon_{\perp}\sqrt{\frac{\varepsilon_{\|}}{\varepsilon_{1}}}C_{-} + \varepsilon_{\perp}\sqrt{\frac{\varepsilon_{\|}}{\varepsilon_{1}}}D_{-} \quad &+ \quad \varepsilon_{\perp}C_{-} + \varepsilon_{\perp}D_{-} = 0 \\ - C_{-}\frac{(\varepsilon_{\perp} - \varepsilon_{\perp}\sqrt{\frac{\varepsilon_{\|}}{\varepsilon_{1}}})}{(\varepsilon_{\perp} + \varepsilon_{\perp}\sqrt{\frac{\varepsilon_{\|}}{\varepsilon_{1}}})} \quad &= \quad D_{-} \end{aligned} \tag{3.284}$$

Substitution of equation (3.285) into equations (3.278) and (3.280) gives

(A)

$$\begin{aligned} &- \quad A_{\|}e^{-jk_{0}d} + B_{\|}e^{jk_{0}d} - \varepsilon_{\perp}\sqrt{\frac{\varepsilon_{\|}}{\varepsilon_{1}}}C_{-}e^{-jk_{-}d} \\ &- \quad \varepsilon_{\perp}C_{-}\frac{(\sqrt{\frac{\varepsilon_{\|}}{\varepsilon_{1}}}\varepsilon_{\perp} - \varepsilon_{\perp}\sqrt{\frac{\varepsilon_{\|}}{\varepsilon_{1}}}\sqrt{\frac{\varepsilon_{\|}}{\varepsilon_{1}}})}{(\varepsilon_{\perp} + \varepsilon_{\perp}\sqrt{\frac{\varepsilon_{\|}}{\varepsilon_{1}}})}e^{jk_{-}d} = 0 \end{aligned}$$

$$\begin{aligned} &- \quad A_{\|}e^{-jk_{0}d} + B_{\|}e^{jk_{0}d} + \{-\varepsilon_{\perp}\sqrt{\frac{\varepsilon_{\|}}{\varepsilon_{1}}}e^{-jk_{-}d} \\ &- \quad \varepsilon_{\perp}\frac{(\sqrt{\frac{\varepsilon_{\|}}{\varepsilon_{1}}}\varepsilon_{\perp} - \varepsilon_{\perp}\sqrt{\frac{\varepsilon_{\|}}{\varepsilon_{1}}}\sqrt{\frac{\varepsilon_{\|}}{\varepsilon_{1}}})}{(\varepsilon_{\perp} + \varepsilon_{\perp}\sqrt{\frac{\varepsilon_{\|}}{\varepsilon_{1}}})}e^{jk_{-}d}\}C_{-} = 0 \end{aligned}$$

And

(B)

$$
\begin{aligned}
- &\quad A_{\|}e^{-jk_0 d} - B_{\|}e^{jk_0 d} - \varepsilon_{\perp}C_{-}e^{-jk_{-}d} \\
+ &\quad C_{-}\frac{(\varepsilon_{\perp}\varepsilon_{\perp} - \varepsilon_{\perp}^{2}\sqrt{\frac{\varepsilon_{\|}}{\varepsilon_1}})}{(\varepsilon_{\perp} + \varepsilon_{\perp}\sqrt{\frac{\varepsilon_{\|}}{\varepsilon_1}})}e^{jk_{-}d} = 0 \\
- &\quad A_{\|}e^{-jk_0 d} - B_{\|}e^{jk_0 d} + \{-\varepsilon_{\perp}e^{-jk_{-}d} \\
+ &\quad \frac{(\varepsilon_{\perp}\varepsilon_{\perp} - \varepsilon_{\perp}^{2}\sqrt{\frac{\varepsilon_{\|}}{\varepsilon_1}})}{(\varepsilon_{\perp} + \varepsilon_{\perp}\sqrt{\frac{\varepsilon_{\|}}{\varepsilon_1}})}e^{jk_{-}d}\}C_{-} = 0
\end{aligned}
\qquad (3.285)
$$

Removing C_- from both of (A) and (B) in equations (3.285), we get

$$
\begin{aligned}
- & \quad A_{\parallel} e^{-jk_0 d}\{-\varepsilon_{\perp} e^{-jk_- d} + \frac{(\varepsilon_{\perp}\varepsilon_{\perp} - \varepsilon_{\perp}^2\sqrt{\frac{\varepsilon_{\parallel}}{\varepsilon_1}})}{(\varepsilon_{\perp} + \varepsilon_{\perp}\sqrt{\frac{\varepsilon_{\parallel}}{\varepsilon_1}})} e^{jk_- d}\} \\
+ & \quad B_{\parallel} e^{jk_0 d}\{-\varepsilon_{\perp} e^{-jk_- d} + \frac{(\varepsilon_{\perp}\varepsilon_{\perp} - \varepsilon_{\perp}^2\sqrt{\frac{\varepsilon_{\parallel}}{\varepsilon_1}})}{(\varepsilon_{\perp} + \varepsilon_{\perp}\sqrt{\frac{\varepsilon_{\parallel}}{\varepsilon_1}})} e^{jk_- d}\} \\
+ & \quad \{-\varepsilon_{\perp}\sqrt{\frac{\varepsilon_{\parallel}}{\varepsilon_1}} e^{-jk_- d} - \varepsilon_{\perp} \frac{(\sqrt{\frac{\varepsilon_{\parallel}}{\varepsilon_1}}\varepsilon_{\perp} - \varepsilon_{\perp}\sqrt{\frac{\varepsilon_{\parallel}}{\varepsilon_1}}\sqrt{\frac{\varepsilon_{\parallel}}{\varepsilon_1}})}{(\varepsilon_{\perp} + \varepsilon_{\perp}\sqrt{\frac{\varepsilon_{\parallel}}{\varepsilon_1}})} e^{jk_- d}\} \\
\star & \quad \{-\varepsilon_{\perp} e^{-jk_- d} + \frac{(\varepsilon_{\perp}\varepsilon_{\perp} - \varepsilon_{\perp}^2\sqrt{\frac{\varepsilon_{\parallel}}{\varepsilon_1}})}{(\varepsilon_{\perp} + \varepsilon_{\perp}\sqrt{\frac{\varepsilon_{\parallel}}{\varepsilon_1}})} e^{jk_- d}\} C_- \\
= & \quad 0
\end{aligned}
$$

and

$$
\begin{aligned}
& A_{\|} e^{-jk_0 d}\{-\varepsilon_\perp \sqrt{\frac{\varepsilon_\|}{\varepsilon_1}} e^{-jk_- d} \\
- & \quad \varepsilon_\perp \frac{(\sqrt{\frac{\varepsilon_\|}{\varepsilon_1}} \varepsilon_\perp - \varepsilon_\perp \sqrt{\frac{\varepsilon_\|}{\varepsilon_1}} \sqrt{\frac{\varepsilon_\|}{\varepsilon_1}})}{(\varepsilon_\perp + \varepsilon_\perp \sqrt{\frac{\varepsilon_\|}{\varepsilon_1}})} e^{jk_- d}\} \\
+ & \quad B_{\|} e^{jk_0 d}\{-\varepsilon_\perp \sqrt{\frac{\varepsilon_\|}{\varepsilon_1}} e^{-jk_- d} \\
- & \quad \varepsilon_\perp \frac{(\sqrt{\frac{\varepsilon_\|}{\varepsilon_1}} \varepsilon_\perp - \varepsilon_\perp \sqrt{\frac{\varepsilon_\|}{\varepsilon_1}} \sqrt{\frac{\varepsilon_\|}{\varepsilon_1}})}{(\varepsilon_\perp + \varepsilon_\perp \sqrt{\frac{\varepsilon_\|}{\varepsilon_1}})} e^{jk_- d}\} \\
- & \quad \{-\varepsilon_\perp e^{-jk_- d} + \frac{(\varepsilon_\perp \varepsilon_\perp - \varepsilon_\perp^2 \sqrt{\frac{\varepsilon_\|}{\varepsilon_1}})}{(\varepsilon_\perp + \varepsilon_\perp \sqrt{\frac{\varepsilon_\|}{\varepsilon_1}})} e^{jk_- d}\} \\
\star & \quad \{-\varepsilon_\perp \sqrt{\frac{\varepsilon_\|}{\varepsilon_1}} e^{-jk_- d} \\
- & \quad \varepsilon_\perp \frac{(\sqrt{\frac{\varepsilon_\|}{\varepsilon_1}} \varepsilon_\perp - \varepsilon_\perp \sqrt{\frac{\varepsilon_\|}{\varepsilon_1}} \sqrt{\frac{\varepsilon_\|}{\varepsilon_1}})}{(\varepsilon_\perp + \varepsilon_\perp \sqrt{\frac{\varepsilon_\|}{\varepsilon_1}})} e^{jk_- d}\} C_- \\
= & \quad 0
\end{aligned}
$$

Then

$$-A_{\parallel}e^{-jk_0d}(\{-\varepsilon_\perp e^{-jk_-d}+\frac{(\varepsilon_\perp\varepsilon_\perp-\varepsilon_\perp^2\sqrt{\frac{\varepsilon_\parallel}{\varepsilon_1}})}{(\varepsilon_\perp+\varepsilon_\perp\sqrt{\frac{\varepsilon_\parallel}{\varepsilon_1}})}e^{jk_-d}\}$$

$$-\{-\varepsilon_\perp\sqrt{\frac{\varepsilon_\parallel}{\varepsilon_1}}e^{-jk_-d}-\varepsilon_\perp\frac{(\sqrt{\frac{\varepsilon_\parallel}{\varepsilon_1}}\varepsilon_\perp-\varepsilon_\perp\sqrt{\frac{\varepsilon_\parallel}{\varepsilon_1}}\sqrt{\frac{\varepsilon_\parallel}{\varepsilon_1}})}{(\varepsilon_\perp+\varepsilon_\perp\sqrt{\frac{\varepsilon_\parallel}{\varepsilon_1}})}e^{jk_-d}\})$$

$$+B_{\parallel}e^{jk_0d}(\{-\varepsilon_\perp e^{-jk_-d}+\frac{(\varepsilon_\perp\varepsilon_\perp-\varepsilon_\perp^2\sqrt{\frac{\varepsilon_\parallel}{\varepsilon_1}})}{(\varepsilon_\perp+\varepsilon_\perp\sqrt{\frac{\varepsilon_\parallel}{\varepsilon_1}})}e^{jk_-d}\}$$

$$+\{-\varepsilon_\perp\sqrt{\frac{\varepsilon_\parallel}{\varepsilon_1}}e^{-jk_-d}-\varepsilon_\perp\frac{(\sqrt{\frac{\varepsilon_\parallel}{\varepsilon_1}}\varepsilon_\perp-\varepsilon_\perp\sqrt{\frac{\varepsilon_\parallel}{\varepsilon_1}}\sqrt{\frac{\varepsilon_\parallel}{\varepsilon_1}})}{(\varepsilon_\perp+\varepsilon_\perp\sqrt{\frac{\varepsilon_\parallel}{\varepsilon_1}})}e^{jk_-d}\})$$

$$=0$$

giving

$$\begin{aligned}
- &\quad A_{\parallel}e^{-jk_0d}(\{-e^{-jk_-d}+\frac{(1-\sqrt{\frac{\varepsilon_\parallel}{\varepsilon_1}})}{(1+\sqrt{\frac{\varepsilon_\parallel}{\varepsilon_1}})}e^{jk_-d}\} \\
- &\quad \{-\sqrt{\frac{\varepsilon_\parallel}{\varepsilon_1}}e^{-jk_-d}-\frac{(\sqrt{\frac{\varepsilon_\parallel}{\varepsilon_1}}-\sqrt{\frac{\varepsilon_\parallel}{\varepsilon_1}}\sqrt{\frac{\varepsilon_\parallel}{\varepsilon_1}})}{(1+\sqrt{\frac{\varepsilon_\parallel}{\varepsilon_1}})}e^{jk_-d}\}) \\
+ &\quad B_{\parallel}e^{jk_0d}(\{-e^{-jk_-d}+\frac{(1-\sqrt{\frac{\varepsilon_\parallel}{\varepsilon_1}})}{(1+\sqrt{\frac{\varepsilon_\parallel}{\varepsilon_1}})}e^{jk_-d}\} \\
+ &\quad \{-\sqrt{\frac{\varepsilon_\parallel}{\varepsilon_1}}e^{-jk_-d}-\frac{(\sqrt{\frac{\varepsilon_\parallel}{\varepsilon_1}}-\sqrt{\frac{\varepsilon_\parallel}{\varepsilon_1}}\sqrt{\frac{\varepsilon_\parallel}{\varepsilon_1}})}{(1+\sqrt{\frac{\varepsilon_\parallel}{\varepsilon_1}})}e^{jk_-d}\}) \\
= &\quad 0
\end{aligned}$$

where

$$\frac{B_{\|} e^{2jk_0 d}}{A_{\|}} = \Gamma_{22} \tag{3.286}$$

From equations (3.281) and (3.283) we have

$$D_+ = \frac{-[1 - \sqrt{\frac{\varepsilon_\perp}{\varepsilon_1}}]}{[1 + \sqrt{\frac{\varepsilon_\perp}{\varepsilon_1}}]} C_+ \tag{3.287}$$

Substituting equation (3.288) into equations (3.277) and (3.279), we have

$$\begin{aligned} A_\perp & e^{-jk_0 d} + B_\perp e^{jk_0 d} + \{-e^{-jk_+ d} \\ + & \frac{[1 - \sqrt{\frac{\varepsilon_\perp}{\varepsilon_1}}]}{[1 + \sqrt{\frac{\varepsilon_\perp}{\varepsilon_1}}]} e^{jk_+ d}\} C_+ = 0 \end{aligned} \tag{3.288}$$

and

$$\begin{aligned} & A_\perp e^{-jk_0 d} - B_\perp e^{jk_0 d} + \\ & \{-\sqrt{\frac{\varepsilon_\perp}{\varepsilon_1}} e^{-jk_+ d} \\ & -\sqrt{\frac{\varepsilon_\perp}{\varepsilon_1}} \star \frac{[1 - \sqrt{\frac{\varepsilon_\perp}{\varepsilon_1}}]}{[1 + \sqrt{\frac{\varepsilon_\perp}{\varepsilon_1}}]} e^{jk_+ d}\} C_+ = 0 \end{aligned} \tag{3.289}$$

Preparing to remove C_+ from equations (3.289) and (3.290), we have

$$\begin{aligned} A_\perp & e^{-jk_0 d}\{-\sqrt{\frac{\varepsilon_\perp}{\varepsilon_1}} e^{-jk_+ d} - \sqrt{\frac{\varepsilon_\perp}{\varepsilon_1}} \star \frac{[1 - \sqrt{\frac{\varepsilon_\perp}{\varepsilon_1}}]}{[1 + \sqrt{\frac{\varepsilon_\perp}{\varepsilon_1}}]} e^{jk_+ d}\} \\ + & B_\perp e^{jk_0 d}\{-\sqrt{\frac{\varepsilon_\perp}{\varepsilon_1}} e^{-jk_+ d} - \sqrt{\frac{\varepsilon_\perp}{\varepsilon_1}} \star \frac{[1 - \sqrt{\frac{\varepsilon_\perp}{\varepsilon_1}}]}{[1 + \sqrt{\frac{\varepsilon_\perp}{\varepsilon_1}}]} e^{jk_+ d}\} \\ + & \{-e^{-jk_+ d} + \frac{[1 - \sqrt{\frac{\varepsilon_\perp}{\varepsilon_1}}]}{[1 + \sqrt{\frac{\varepsilon_\perp}{\varepsilon_1}}]} e^{jk_+ d}\} \\ * & \{-\sqrt{\frac{\varepsilon_\perp}{\varepsilon_1}} e^{-jk_+ d} - \sqrt{\frac{\varepsilon_\perp}{\varepsilon_1}} \star \frac{[1 - \sqrt{\frac{\varepsilon_\perp}{\varepsilon_1}}]}{[1 + \sqrt{\frac{\varepsilon_\perp}{\varepsilon_1}}]} e^{jk_+ d}\} C_+ = 0 \end{aligned}$$

and

$$
\begin{aligned}
-A_{\perp} \quad & e^{-jk_0 d}\{-e^{-jk_+ d} + \frac{[1-\sqrt{\frac{\varepsilon_{\perp}}{\varepsilon_1}}]}{[1+\sqrt{\frac{\varepsilon_{\perp}}{\varepsilon_1}}]} e^{jk_+ d}\} \\
+ \quad & B_{\perp} e^{jk_0 d}\{-e^{-jk_+ d} + \frac{[1-\sqrt{\frac{\varepsilon_{\perp}}{\varepsilon_1}}]}{[1+\sqrt{\frac{\varepsilon_{\perp}}{\varepsilon_1}}]} e^{jk_+ d}\} \\
- \quad & \{-\sqrt{\frac{\varepsilon_{\perp}}{\varepsilon_1}} e^{-jk_+ d} - \sqrt{\frac{\varepsilon_{\perp}}{\varepsilon_1}} \star \frac{[1-\sqrt{\frac{\varepsilon_{\perp}}{\varepsilon_1}}]}{[1+\sqrt{\frac{\varepsilon_{\perp}}{\varepsilon_1}}]} e^{jk_+ d}\} \\
* \quad & \{-e^{-jk_+ d} + \frac{[1-\sqrt{\frac{\varepsilon_{\perp}}{\varepsilon_1}}]}{[1+\sqrt{\frac{\varepsilon_{\perp}}{\varepsilon_1}}]} e^{jk_+ d}\} C_+ = 0
\end{aligned}
$$

By combining the two equations above, we get

$$
\begin{aligned}
A_{\perp} \quad & e^{-jk_0 d}(\{-\sqrt{\frac{\varepsilon_{\perp}}{\varepsilon_1}} e^{-jk_+ d} - \sqrt{\frac{\varepsilon_{\perp}}{\varepsilon_1}} \star \frac{[1-\sqrt{\frac{\varepsilon_{\perp}}{\varepsilon_1}}]}{[1+\sqrt{\frac{\varepsilon_{\perp}}{\varepsilon_1}}]} e^{jk_+ d}\} \\
- \quad & \{-e^{-jk_+ d} + \frac{[1-\sqrt{\frac{\varepsilon_{\perp}}{\varepsilon_1}}]}{[1+\sqrt{\frac{\varepsilon_{\perp}}{\varepsilon_1}}]} e^{jk_+ d}\}) \\
+ \quad B_{\perp} \quad & e^{jk_0 d}(\{-\sqrt{\frac{\varepsilon_{\perp}}{\varepsilon_1}} e^{-jk_+ d} - \sqrt{\frac{\varepsilon_{\perp}}{\varepsilon_1}} \star \frac{[1-\sqrt{\frac{\varepsilon_{\perp}}{\varepsilon_1}}]}{[1+\sqrt{\frac{\varepsilon_{\perp}}{\varepsilon_1}}]} e^{jk_+ d}\} \\
+ \quad & \{-e^{-jk_+ d} + \frac{[1-\sqrt{\frac{\varepsilon_{\perp}}{\varepsilon_1}}]}{[1+\sqrt{\frac{\varepsilon_{\perp}}{\varepsilon_1}}]} e^{jk_+ d}\}) = 0
\end{aligned}
$$

where

$$
\frac{B_{\perp} e^{2jk_0 d}}{A_{\perp}} = \Gamma_{11} \tag{3.290}
$$

From equations (3.276) and (3.278), we have

$$\begin{aligned}2A_\perp e^{-jk_0 d} &+ (-e^{-jk_+ d} - \sqrt{\frac{\varepsilon_\perp}{\varepsilon_1}} e^{-jk_+ d})C_+ \\ &+ (\sqrt{\frac{\varepsilon_\perp}{\varepsilon_1}} e^{jk_+ d} - e^{jk_+ d})D_+ \\ &= 0\end{aligned}$$

or

$$-\frac{2A_\perp e^{-jk_0 d} + (-e^{-jk_+ d} - \sqrt{\frac{\varepsilon_\perp}{\varepsilon_1}} e^{-jk_+ d})C_+}{(\sqrt{\frac{\varepsilon_\perp}{\varepsilon_1}} e^{jk_+ d} - e^{jk_+ d})} = D_+ \qquad (3.291)$$

Substituting equation (3.292) into equation (3.281) and (3.283), we get

$$\begin{aligned}&- \quad 2A_\perp \frac{e^{-jk_0 d}}{(\sqrt{\frac{\varepsilon_\perp}{\varepsilon_1}} e^{jk_+ d} - e^{jk_+ d})} \\ &+ \quad [1 - \frac{(-e^{-jk_+ d} - \sqrt{\frac{\varepsilon_\perp}{\varepsilon_1}} e^{-jk_+ d})}{(\sqrt{\frac{\varepsilon_\perp}{\varepsilon_1}} e^{jk_+ d} - e^{jk_+ d})}]C_+ - F_\perp \\ &= \quad 0\end{aligned}$$

And

$$\begin{aligned}& \quad 2A_\perp \frac{e^{-jk_0 d}}{(\sqrt{\frac{\varepsilon_\perp}{\varepsilon_1}} e^{jk_+ d} - e^{jk_+ d})} \\ &+ \quad [1 + \frac{(-e^{-jk_+ d} - \sqrt{\frac{\varepsilon_\perp}{\varepsilon_1}} e^{-jk_+ d})}{(\sqrt{\frac{\varepsilon_\perp}{\varepsilon_1}} e^{jk_+ d} - e^{jk_+ d})}]C_+ - \sqrt{\frac{\varepsilon_1}{\varepsilon_\perp}} F_\perp \\ &= \quad 0\end{aligned}$$

By combining both equations above, we get

$$2A_{\perp}\frac{e^{-jk_0 d}}{(\sqrt{\frac{\varepsilon_{\perp}}{\varepsilon_1}}e^{jk_+ d}-e^{jk_+ d})}\star[1+\frac{(-e^{-jk_+ d}-\sqrt{\frac{\varepsilon_{\perp}}{\varepsilon_1}}e^{-jk_+ d})}{(\sqrt{\frac{\varepsilon_{\perp}}{\varepsilon_1}}e^{jk_+ d}-e^{jk_+ d})}]$$

$$+[1+\frac{(-e^{-jk_+ d}-\sqrt{\frac{\varepsilon_{\perp}}{\varepsilon_1}}e^{-jk_+ d})}{(\sqrt{\frac{\varepsilon_{\perp}}{\varepsilon_1}}e^{jk_+ d}-e^{jk_+ d})}]F_{\perp}$$

$$+2A_{\perp}\frac{e^{-jk_0 d}}{(\sqrt{\frac{\varepsilon_{\perp}}{\varepsilon_1}}e^{jk_+ d}-e^{jk_+ d})}\star[1-\frac{(-e^{-jk_+ d}-\sqrt{\frac{\varepsilon_{\perp}}{\varepsilon_1}}e^{-jk_+ d})}{(\sqrt{\frac{\varepsilon_{\perp}}{\varepsilon_1}}e^{jk_+ d}-e^{jk_+ d})}]$$

$$-\sqrt{\frac{\varepsilon_1}{\varepsilon_{\perp}}}\star[1-\frac{(-e^{-jk_+ d}-\sqrt{\frac{\varepsilon_{\perp}}{\varepsilon_1}}e^{-jk_+ d})}{(\sqrt{\frac{\varepsilon_{\perp}}{\varepsilon_1}}e^{jk_+ d}-e^{jk_+ d})}]F_{\perp}=0$$

Or

$$4A_{\perp}e^{-jk_0 d}+\{(\sqrt{\frac{\varepsilon_{\perp}}{\varepsilon_1}}e^{jk_+ d}-e^{jk_+ d}-e^{-jk_+ d}-\sqrt{\frac{\varepsilon_{\perp}}{\varepsilon_1}}e^{-jk_+ d})\}F_{\perp}$$

$$-\sqrt{\frac{\varepsilon_1}{\varepsilon_{\perp}}}\star\{(\sqrt{\frac{\varepsilon_{\perp}}{\varepsilon_1}}e^{jk_+ d}-e^{jk_+ d}+e^{-jk_+ d}+\sqrt{\frac{\varepsilon_{\perp}}{\varepsilon_1}}e^{-jk_+ d})\}F_{\perp}=0$$

Hence,

$$\frac{-4A_{\perp}e^{-jk_0 d}}{[2j(\sqrt{\frac{\varepsilon_{\perp}}{\varepsilon_1}}+\sqrt{\frac{\varepsilon_1}{\varepsilon_{\perp}}})\sin(k_+ d)-2\cos(k_+ d)]}=F_{\perp}$$

where

$$\frac{F_{\perp}e^{jk_0 d}}{A_{\perp}}=T_{11} \qquad (3.292)$$

From equations (3.277) and (3.279), we have

$$-2A_{\parallel}e^{-jk_0 d}+(-\varepsilon_{\perp}\sqrt{\frac{\varepsilon_{\parallel}}{\varepsilon_1}}e^{-jk_- d}-\varepsilon_{\perp}e^{-jk_- d})C_-$$

$$+(\varepsilon_{\perp}\sqrt{\frac{\varepsilon_{\parallel}}{\varepsilon_1}}e^{jk_- d}-\varepsilon_{\perp}e^{jk_- d})D_-=0$$

Or

$$2A_{\|}\frac{e^{-jk_0d}}{(\varepsilon_{\perp}\sqrt{\frac{\varepsilon_{\|}}{\varepsilon_1}}e^{jk_-d}-\varepsilon_{\perp}e^{jk_-d})}$$

$$-\frac{(-\varepsilon_{\perp}\sqrt{\frac{\varepsilon_{\|}}{\varepsilon_1}}e^{-jk_-d}-\varepsilon_{\perp}e^{-jk_-d})}{(\varepsilon_{\perp}\sqrt{\frac{\varepsilon_{\|}}{\varepsilon_1}}e^{jk_-d}-\varepsilon_{\perp}e^{jk_-d})}C_-=D_- \qquad (3.293)$$

Substituting equation (3.293) into equations (3.281) and (3.283), we get

$$F_{\|}+2\varepsilon_{\perp}A_{\|}\frac{e^{-jk_0d}}{(\varepsilon_{\perp}\sqrt{\frac{\varepsilon_{\|}}{\varepsilon_1}}e^{jk_-d}-\varepsilon_{\perp}e^{jk_-d})}$$

$$+\{\frac{(\varepsilon_{\perp}\sqrt{\frac{\varepsilon_{\|}}{\varepsilon_1}}e^{-jk_-d}+\varepsilon_{\perp}e^{-jk_-d})}{(\varepsilon_{\perp}\sqrt{\frac{\varepsilon_{\|}}{\varepsilon_1}}e^{jk_-d}-\varepsilon_{\perp}e^{jk_-d})}\varepsilon_{\perp}+\varepsilon_{\perp}\}C_-=0$$

And

$$-\quad 2A_{\|}\frac{e^{-jk_0d}\varepsilon_{\perp}}{(\varepsilon_{\perp}\sqrt{\frac{\varepsilon_{\|}}{\varepsilon_1}}e^{jk_-d}-\varepsilon_{\perp}e^{jk_-d})}$$

$$+\quad \{1-\frac{(\varepsilon_{\perp}\sqrt{\frac{\varepsilon_{\|}}{\varepsilon_1}}e^{-jk_-d}+\varepsilon_{\perp}e^{-jk_-d})}{(\varepsilon_{\perp}\sqrt{\frac{\varepsilon_{\|}}{\varepsilon_1}}e^{jk_-d}-\varepsilon_{\perp}e^{jk_-d})}\}\varepsilon_{\perp}C_-+\sqrt{\frac{\varepsilon_1}{\varepsilon_{\|}}}F_{\|}=0$$

Combining both equations results in

$$\varepsilon_{\perp}F_{\|}\{1-\frac{(\varepsilon_{\perp}\sqrt{\frac{\varepsilon_{\|}}{\varepsilon_1}}e^{-jk_-d}+\varepsilon_{\perp}e^{-jk_-d})}{(\varepsilon_{\perp}\sqrt{\frac{\varepsilon_{\|}}{\varepsilon_1}}e^{jk_-d}-\varepsilon_{\perp}e^{jk_-d})}\}$$

$$-\sqrt{\frac{\varepsilon_1}{\varepsilon_{\|}}}\{\frac{(\varepsilon_{\perp}\sqrt{\frac{\varepsilon_{\|}}{\varepsilon_1}}e^{-jk_-d}+\varepsilon_{\perp}e^{-jk_-d})}{(\varepsilon_{\perp}\sqrt{\frac{\varepsilon_{\|}}{\varepsilon_1}}e^{jk_-d}-\varepsilon_{\perp}e^{jk_-d})}\varepsilon_{\perp}+\varepsilon_{\perp}\}F_{\|}$$

$$+4\varepsilon_{\perp}^2A_{\|}\frac{e^{-jk_0d}}{(\varepsilon_{\perp}\sqrt{\frac{\varepsilon_{\|}}{\varepsilon_1}}e^{jk_-d}-\varepsilon_{\perp}e^{jk_-d})}=0$$

Hence,

$$
\begin{aligned}
(\quad - & \quad \{1 - \frac{(\varepsilon_\perp \sqrt{\frac{\varepsilon_\parallel}{\varepsilon_1}} e^{-jk_-d} + \varepsilon_\perp e^{-jk_-d})}{(\varepsilon_\perp \sqrt{\frac{\varepsilon_\parallel}{\varepsilon_1}} e^{jk_-d} - \varepsilon_\perp e^{jk_-d})}\}\varepsilon_\perp \\
+ & \quad \sqrt{\frac{\varepsilon_1}{\varepsilon_\parallel}}\{\frac{(\varepsilon_\perp \sqrt{\frac{\varepsilon_\parallel}{\varepsilon_1}} e^{-jk_-d} + \varepsilon_\perp e^{-jk_-d})}{(\varepsilon_\perp \sqrt{\frac{\varepsilon_\parallel}{\varepsilon_1}} e^{jk_-d} - \varepsilon_\perp e^{jk_-d})}\varepsilon_\perp + \varepsilon_\perp\})F_\parallel \\
= & \quad 4\varepsilon_\perp^2 A_\parallel \frac{e^{-jk_0 d}}{(\varepsilon_\perp \sqrt{\frac{\varepsilon_\parallel}{\varepsilon_1}} e^{jk_-d} - \varepsilon_\perp e^{jk_-d})}
\end{aligned}
$$

giving

$$
\begin{aligned}
(\quad - & \quad \{1 - \frac{(\sqrt{\frac{\varepsilon_\parallel}{\varepsilon_1}} e^{-jk_-d} + e^{-jk_-d})}{(\sqrt{\frac{\varepsilon_\parallel}{\varepsilon_1}} e^{jk_-d} - e^{jk_-d})}\} \\
+ & \quad \sqrt{\frac{\varepsilon_1}{\varepsilon_\parallel}}\{\frac{(\sqrt{\frac{\varepsilon_\parallel}{\varepsilon_1}} e^{-jk_-d} + e^{-jk_-d})}{(\sqrt{\frac{\varepsilon_\parallel}{\varepsilon_1}} e^{jk_-d} - e^{jk_-d})} + 1\})F_\parallel \\
= & \quad 4A_\parallel \frac{e^{-jk_0 d}}{(\sqrt{\frac{\varepsilon_\parallel}{\varepsilon_1}} e^{jk_-d} - e^{jk_-d})}
\end{aligned}
$$

where

$$
\frac{F_\parallel e^{jk_0 d}}{A_\parallel} = T_{22} \tag{3.294}
$$

We see Figures 3.31 below for the numerical calculations of the parallel transmitted and reflected components of the waves (see Figure 3.32 as well); and by applying FFT (Fast Fourier Transform) tool in Matlab, I was able to transform the analytical former solutions to signals in time domain after multiplying them with the frequency domain equation of the raised cosine pulse of rank 2.[18]

[18] See Appendix A.

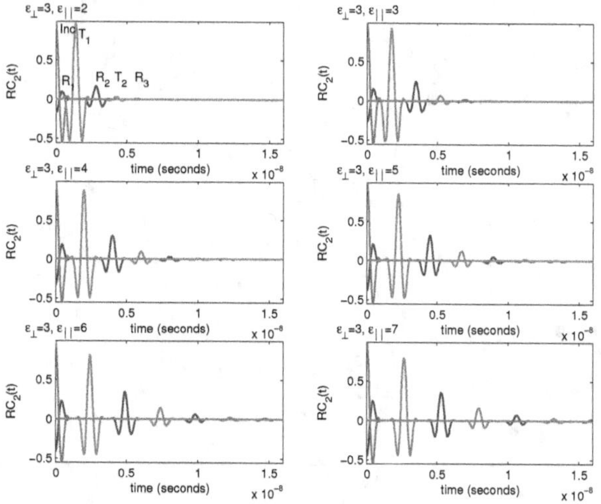

Figure 3.31.: Time Domain Representation of the Parallel Components of the Transmitted and Reflected Waves of RC_2 Pulses through a Uniaxial Plate, When f=1 GHz, d=0.3 m, $\varepsilon_\perp$=3 , $\varepsilon_\parallel$=2 to 7.

Looking into the first figure of the set above, we notice that: Frequency= 1 GHz, $\varepsilon_\perp = 3$, $\varepsilon_\parallel = 2$, $\varepsilon_1 = 1$, we get from equation (3.271) the phase velocity for each wave vector $v_{p+} = \frac{\omega}{k_+} = \frac{c_0}{\sqrt{\varepsilon_\perp}} = \frac{3*10^8}{\sqrt{3}} = 1.732 \star 10^8$ m/sec, and $v_{p-} = \frac{\omega}{k_-} = \frac{c_0}{\sqrt{\varepsilon_\parallel}} = \frac{3*10^8}{\sqrt{2}} = 2.121 \star 10^8$ m/sec.

And checking out the time in which the first transmitted signal has appeared, we notice that it equals to $1.414 * 10^{-9}$ seconds, and dividing the thickness of the plate (d=0.3 meter) by it would give the phase velocity within the uniaxial medium in which the transmitted wave has been carried all along and that equals to $2.121 * 10^8$ m/sec, which is the same speed of the extraordinary wave.

If $\varepsilon_\parallel$ is to be varied while $\varepsilon_\perp$ is fixed, then, Figures 3.31 show that the time in which the signal is transmitted varies with the changing in value of the parameter $\varepsilon_{||}$ which corresponds to the extraordinary wave. Hence, varying the $\varepsilon_{\perp}$ component won't have an effect on equations (3.286) and (3.294).

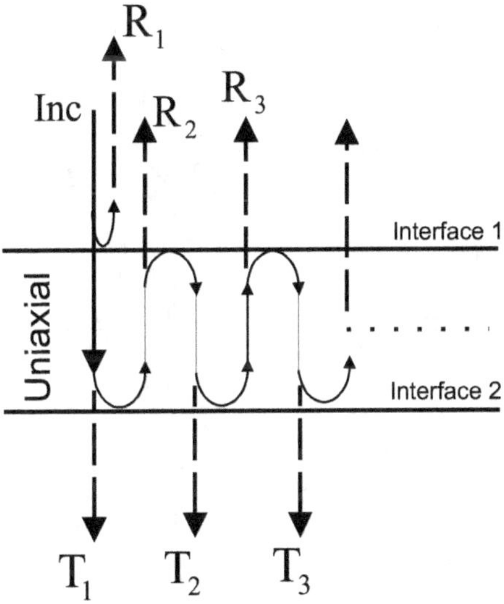

Figure 3.32.: Normal Incidence of RC_2 Pulse on a Uniaxial Plate Showing the Direction of the Resulted Reflected and Transmitted Waves.

Another remark to be noticed is that the time period between each two consecutive reflected signals or two consecutive transmitted signals equals to the double amount of time in which the phase velocity of the signal allows it to get back to one of the two interfaces after bouncing off the other; and that equals to 2d/velocity.

In contrast, the numerical calculations of the perpendicular transmitted and reflected components of the waves are dependent on $\varepsilon_\perp$ rather than on $\varepsilon_\parallel$ as shown in Figures 3.33 below

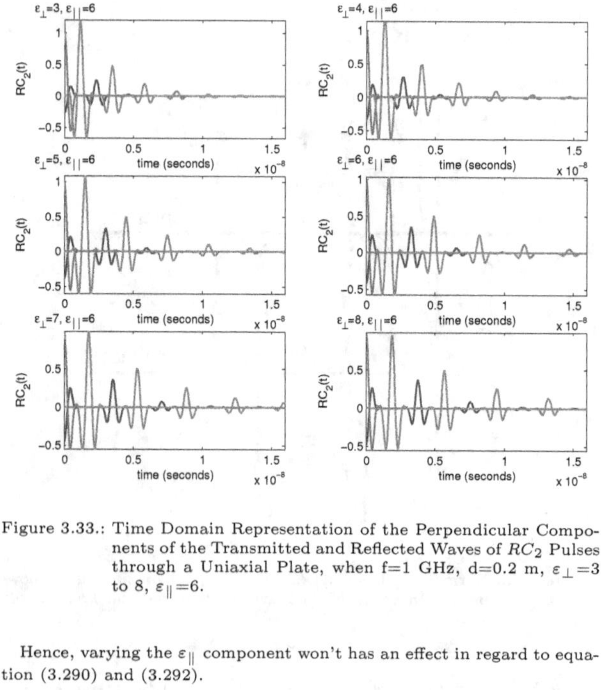

Figure 3.33.: Time Domain Representation of the Perpendicular Components of the Transmitted and Reflected Waves of RC_2 Pulses through a Uniaxial Plate, when f=1 GHz, d=0.2 m, $\varepsilon_\perp$=3 to 8, $\varepsilon_\parallel$=6.

Hence, varying the $\varepsilon_\parallel$ component won't has an effect in regard to equation (3.290) and (3.292).

3.8.2. Optic Axis Parallel to the Normal of the Interface

For a uniform plane wave propagating in isotropic medium and going through an anisotropic uniaxial plate perpendicular to its interface and parallel to the optic axis, and then out of this plate back to the isotropic medium is shown in Figure 3.34 below.

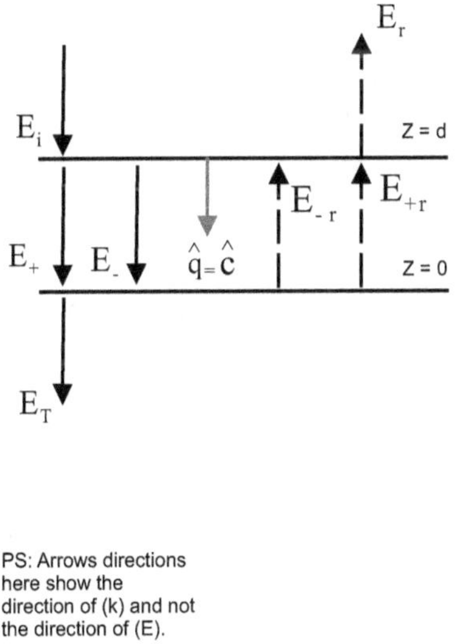

Figure 3.34.: Uniform Plane Wave Propagating through an Uniaxial Plate Parallel to its Optic Axis.

This is the case of equations (3.255). Following the derivation of normal incidence on the interface of two isotropic mediums we have

$$\Gamma = \frac{\eta_d - \eta_0 + (\eta_0 - \eta_d)e^{2jk_d d}}{\eta_0 + \eta_d + \frac{(\eta_0 - \eta_d)^2}{\eta_0 + \eta_d} e^{2jk_d d}} \tag{3.295}$$

where

$$\eta_d = \eta_t = \sqrt{\frac{1}{\varepsilon_\perp}}$$
$$\eta_0 = \sqrt{\frac{1}{\varepsilon_1}}$$

and

$$T = (\frac{2}{1+\frac{\eta_d}{\eta_0}})\frac{(\eta_d+\eta_0)+(\Gamma)(-\eta_d+\eta_0)}{2\eta_0 e^{-jk_d d}} \tag{3.296}$$

3.8.3. Optic Axis with an Angle of 45 Degrees to the Normal of the Interface

Our field vectors will become

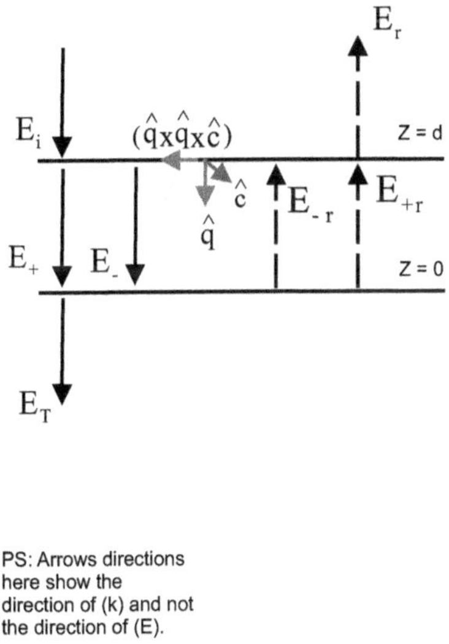

Figure 3.35.: Uniform Plane Wave Propagating through an Uniaxial Plate When the Optic Axis is Having a 45° Angle with the Normal of the Interface.

$$\begin{aligned}
\mathbf{E}_i &= A_\perp(\widehat{\mathbf{q}} \times \widehat{\mathbf{c}})e^{-jk_0 z} \\
&+ A_\parallel[\widehat{\mathbf{q}} \times (\widehat{\mathbf{q}} \times \widehat{\mathbf{c}})]e^{-jk_0 z} && (3.297) \\
\mathbf{E}_r &= B_\perp(\widehat{\mathbf{q}} \times \widehat{\mathbf{c}})e^{jk_0 z} \\
&+ B_\parallel[\widehat{\mathbf{q}} \times (\widehat{\mathbf{q}} \times \widehat{\mathbf{c}})]e^{jk_0 z} && (3.298) \\
\mathbf{E}_+ &= C_+(\widehat{\mathbf{q}} \times \widehat{\mathbf{c}})e^{-jk_+ z} && (3.299) \\
\mathbf{E}_{+r} &= D_+(\widehat{\mathbf{q}} \times \widehat{\mathbf{c}})e^{jk_+ z} && (3.300) \\
\mathbf{E}_- &= [\varepsilon_\perp \widehat{\mathbf{c}} - \frac{k_-^2}{k_0^2}\cos\beta\ \widehat{\mathbf{q}}]C_- e^{-jk_- z} && (3.301) \\
\mathbf{E}_{-r} &= [\varepsilon_\perp \widehat{\mathbf{c}} - \frac{k_-^2}{k_0^2}\cos\beta\ \widehat{\mathbf{q}}]D_- e^{jk_- z} && (3.302) \\
\mathbf{E}_T &= F_\perp(\widehat{\mathbf{q}} \times \widehat{\mathbf{c}})e^{-jk_0 z} \\
&+ F_\parallel[\widehat{\mathbf{q}} \times (\widehat{\mathbf{q}} \times \widehat{\mathbf{c}})]e^{-jk_0 z} && (3.303)
\end{aligned}$$

And

$$\begin{aligned}
\mathbf{H}_i &= \frac{1}{\omega\mu_0}(\mathbf{k}_i \times \mathbf{E}_i) \\
&= \sqrt{\frac{\varepsilon_0\varepsilon_1}{\mu_0}}\{A_\perp[\widehat{\mathbf{q}} \times (\widehat{\mathbf{q}} \times \widehat{\mathbf{c}})]e^{-jk_0 z} \\
&\quad -A_\parallel(\widehat{\mathbf{q}} \times \widehat{\mathbf{c}})e^{-jk_0 z}\} && (3.304) \\
\mathbf{H}_r &= \frac{1}{\omega\mu_0}(\mathbf{k}_r \times \mathbf{E}_r) \\
&= \sqrt{\frac{\varepsilon_0\varepsilon_1}{\mu_0}}\{-B_\perp[\widehat{\mathbf{q}} \times (\widehat{\mathbf{q}} \times \widehat{\mathbf{c}})]e^{jk_0 z} \\
&\quad +B_\parallel(\widehat{\mathbf{q}} \times \widehat{\mathbf{c}})e^{jk_0 z}\} && (3.305)
\end{aligned}$$

$$\begin{aligned}
\mathbf{H}_+ &= \frac{1}{\omega\mu_0}(\mathbf{k}_+ \times \mathbf{E}_+) \\
&= \sqrt{\frac{\varepsilon_0\varepsilon_\perp}{\mu_0}} C_+[\widehat{\mathbf{q}} \times (\widehat{\mathbf{q}} \times \widehat{\mathbf{c}})] e^{-jk_+ z} && (3.306) \\
\mathbf{H}_{+r} &= \frac{1}{\omega\mu_0}(\mathbf{k}_{+r} \times \mathbf{E}_{+r}) \\
&= -\sqrt{\frac{\varepsilon_0\varepsilon_\perp}{\mu_0}} D_+[\widehat{\mathbf{q}} \times (\widehat{\mathbf{q}} \times \widehat{\mathbf{c}})] e^{jk_+ z} && (3.307) \\
\mathbf{H}_- &= \frac{1}{\omega\mu_0}(\mathbf{k}_- \times \mathbf{E}_-) \\
&= \varepsilon_\perp \sqrt{\frac{\varepsilon_0}{\mu_0}} \frac{k_-}{k_0} \; C_-(\widehat{\mathbf{q}} \times \widehat{\mathbf{c}}) e^{-jk_- z} && (3.308) \\
\mathbf{H}_{-r} &= \frac{1}{\omega\mu_0}(\mathbf{k}_- \times \mathbf{E}_-) \\
&= -\varepsilon_\perp \sqrt{\frac{\varepsilon_0}{\mu_0}} \frac{k_-}{k_0} \; D_-(\widehat{\mathbf{q}} \times \widehat{\mathbf{c}}) e^{jk_- z} && (3.309) \\
\mathbf{H}_T &= \frac{1}{\omega\mu_0}(\mathbf{k}_T \times \mathbf{E}_T) \\
&= \sqrt{\frac{\varepsilon_0\varepsilon_1}{\mu_0}} \{F_\perp [\widehat{\mathbf{q}} \times (\widehat{\mathbf{q}} \times \widehat{\mathbf{c}})] e^{-jk_0 z} \\
&\quad - F_\parallel (\widehat{\mathbf{q}} \times \widehat{\mathbf{c}}) e^{-jk_0 z}\} && (3.310)
\end{aligned}$$

with the wave vectors

$$\mathbf{k}_i = k_i\widehat{\mathbf{q}} = -\mathbf{k}_r, \qquad \mathbf{k}_+ = k_+\widehat{\mathbf{q}}, \qquad \mathbf{k}_- = k_-\widehat{\mathbf{q}}$$
$$\mathbf{k}_{+r} = -k_+\widehat{\mathbf{q}}, \qquad \mathbf{k}_{-r} = -k_-\widehat{\mathbf{q}} \tag{3.311}$$

where

$$\begin{aligned}
k_i &= k_0\sqrt{\varepsilon_1} \\
k_+ &= k_0\sqrt{\varepsilon_\perp} \\
k_- &= k_0\sqrt{\frac{\varepsilon_\perp\varepsilon_\parallel}{\widehat{\mathbf{q}} \cdot \overline{\varepsilon} \cdot \widehat{\mathbf{q}}}} \\
&= k_0\sqrt{\frac{\varepsilon_\parallel\varepsilon_\perp}{\varepsilon_\perp + \cos^2\beta\ (\varepsilon_\parallel - \varepsilon_\perp)}} && (3.312)
\end{aligned}$$

Figure 3.36 below shows the relation between the different permittivities in $\frac{k_-}{k_0}$ in equation (3.312) when $\beta = 45^o$

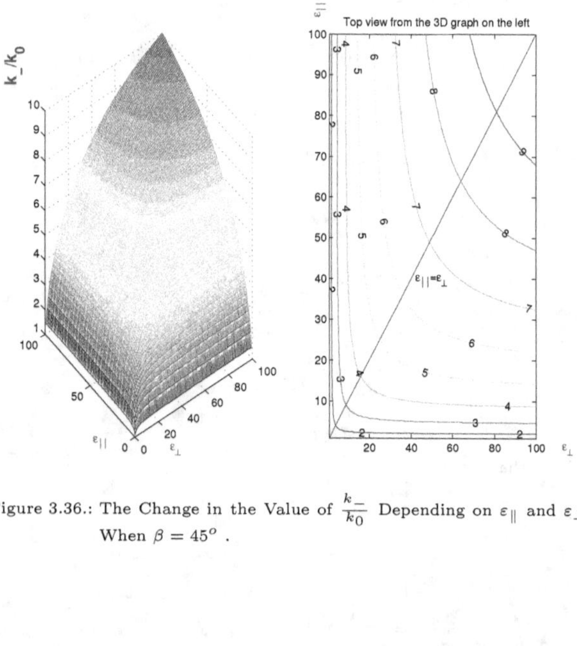

Figure 3.36.: The Change in the Value of $\frac{k_-}{k_0}$ Depending on $\varepsilon_{\parallel}$ and $\varepsilon_{\perp}$ When $\beta = 45^o$.

As we see from the figure above, increasing both values of the perpendicular and parallel permittivities simultaneously gives a higher value for $\frac{k_-}{k_0}$. This affects the outcome of the numerical calculations only for the parallel transmitted and reflected components of the waves, while perpendicular transmitted and reflected components of the waves remain unchanged.

$$
\begin{aligned}
- \quad & A_{\parallel} e^{-jk_0 d}(\{-e^{-jk_- d} + \frac{(1 - \frac{k_-}{k_0}\sqrt{\frac{1}{\varepsilon_1}})}{(1 + \frac{k_-}{k_0}\sqrt{\frac{1}{\varepsilon_1}})} e^{jk_- d}\} \\
- \quad & \{-\frac{k_-}{k_0}\sqrt{\frac{1}{\varepsilon_1}} e^{-jk_- d} - \frac{k_-}{k_0} \frac{(\sqrt{\frac{1}{\varepsilon_1}} - \frac{k_-}{k_0}\sqrt{\frac{1}{\varepsilon_1}}\sqrt{\frac{1}{\varepsilon_1}})}{(1 + \frac{k_-}{k_0}\sqrt{\frac{1}{\varepsilon_1}})} e^{jk_- d}\}) \\
+ \quad & B_{\parallel} e^{jk_0 d}(\{-e^{-jk_- d} + \frac{(1 - \frac{k_-}{k_0}\sqrt{\frac{1}{\varepsilon_1}})}{(1 + \frac{k_-}{k_0}\sqrt{\frac{1}{\varepsilon_1}})} e^{jk_- d}\} \\
+ \quad & \{-\frac{k_-}{k_0}\sqrt{\frac{1}{\varepsilon_1}} e^{-jk_- d} - \frac{k_-}{k_0} \frac{(\sqrt{\frac{1}{\varepsilon_1}} - \frac{k_-}{k_0}\sqrt{\frac{1}{\varepsilon_1}}\sqrt{\frac{1}{\varepsilon_1}})}{(1 + \frac{k_-}{k_0}\sqrt{\frac{1}{\varepsilon_1}})} e^{jk_- d}\}) \\
= \quad & 0
\end{aligned}
$$

where

$$
\frac{B_{\parallel} e^{2jk_0 d}}{A_{\parallel}} = \Gamma_{22} \tag{3.313}
$$

And

$$
\begin{aligned}
& (-\{1 - \frac{(\frac{k_-}{k_0}\sqrt{\frac{1}{\varepsilon_1}} e^{-jk_- d} + e^{-jk_- d})}{(\frac{k_-}{k_0}\sqrt{\frac{1}{\varepsilon_1}} e^{jk_- d} - e^{jk_- d})}\} \frac{k_-}{k_0} \\
+ \quad & \sqrt{\varepsilon_1}\{\frac{(\frac{k_-}{k_0}\sqrt{\frac{1}{\varepsilon_1}} e^{-jk_- d} + e^{-jk_- d})}{(\frac{k_-}{k_0}\sqrt{\frac{1}{\varepsilon_1}} e^{jk_- d} - e^{jk_- d})} + 1\})F_{\parallel} \\
= \quad & 4\frac{k_-}{k_0} A_{\parallel} \frac{e^{-jk_0 d}}{(\frac{k_-}{k_0}\sqrt{\frac{1}{\varepsilon_1}} e^{jk_- d} - e^{jk_- d})}
\end{aligned}
$$

where

$$
\frac{F_{\parallel} e^{jk_0 d}}{A_{\parallel}} = T_{22} \tag{3.314}
$$

Both new equations attained above are not purely perpendicular permittivity dependent or parallel permittivity dependent as before.

Figures 3.37 below show that the tilted axis case will show same behavior of the non-tilted case if the difference between both permittivities is almost zero or small enough so that the apex of one of the curves in the top view within Figure 3.36 will be reached. And deviating from this apex will cause a change directly unless the difference between both permittivities will be kept on the curved line on the same value.

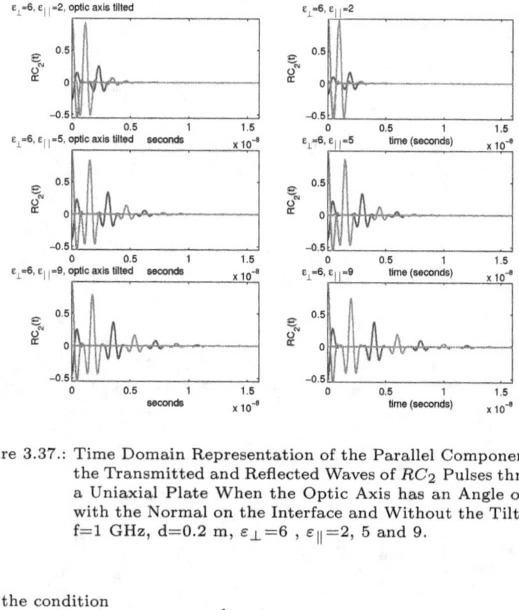

Figure 3.37.: Time Domain Representation of the Parallel Components of the Transmitted and Reflected Waves of RC_2 Pulses through a Uniaxial Plate When the Optic Axis has an Angle of 45° with the Normal on the Interface and Without the Tilt, and f=1 GHz, d=0.2 m, $\varepsilon_{\perp}$=6 , $\varepsilon_{\|}$=2, 5 and 9.

If the condition

$$\frac{k_-}{k_0} \approxeq \sqrt{\varepsilon_{\|}} \tag{3.315}$$

is reached, then the uniaxial plate with tilted angle will react exactly the same as if it were not having this tilt in its optic axis, and rather by having a 90 degrees angle between the optic axis and the normal to the interface. Hence, equations (3.314) and (3.313) will equal equations

(3.294) and (3.286) respectively.

The figures below show the differences among several cases.

Figures 3.38 below shows that when increasing the value of $\varepsilon_{\perp}$ and maintaining the value of $\varepsilon_{\parallel}$ fixed, the wave's travel time within the uniaxial plate will change according to the change of $\varepsilon_{\parallel}$ and $\varepsilon_{\perp}$.

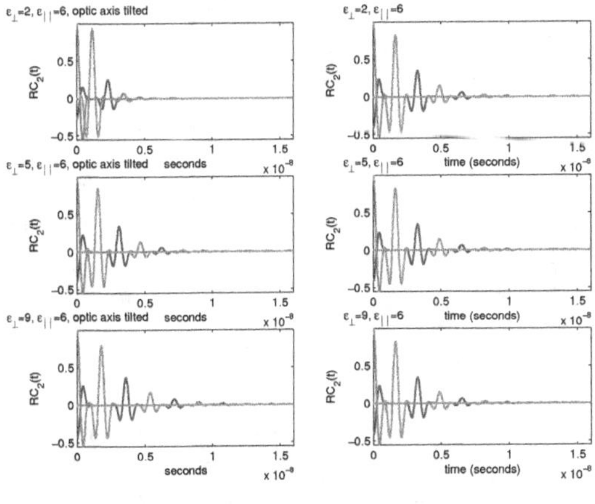

Figure 3.38.: Time Domain Representation of the Parallel Components of the Transmitted and Reflected Waves of RC_2 Pulses through a uniaxial Plate When the optic axis has an Angle of 45° with the Normal on the Interface and Without the Tilt, and f=1 GHz, d=0.2 m, $\varepsilon_{\perp}$=2, 5 and 9 , $\varepsilon_{\parallel}$=6.

In Figures 3.39 below, both perpendicular and parallel permittivities are equal, which gives same results between tilted and non-tilted optic axis polarization.

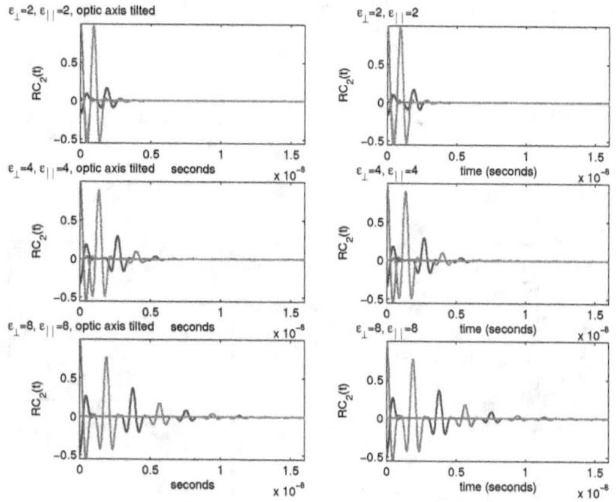

Figure 3.39.: Time Domain Representation of the Parallel Components of the Transmitted and Reflected Waves of RC_2 Pulses through a uniaxial Plate When the Optic Axis has an Angle of 45° with the Normal on the Interface and Without the Tilt, and f=1 GHz, d=0.2 m, $\varepsilon_{\perp}=\varepsilon_{\|}=2$, 4 and 8.

Appendix

A

Raised Cosine Pulse

A.1. Mathematical Functions

The function of a Raised Cosine pulse of rank 2 is

$$f_{RC_2}(t) = \begin{cases} (1 + \cos(\pi f_0 t)) \cos(2\pi f_0 t) & , -T_0 < t < T_0 \\ \\ 0 & , elsewhere \end{cases} \tag{A.1}$$

And to find out the frequency domain equation of A.1, we have

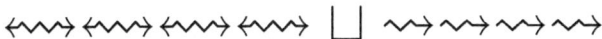

$$
\begin{aligned}
f_{RC_2}(\omega) &= \int_{-T_0}^{T_0} (1 + \frac{e^{j\pi f_0 t}}{2} + \frac{e^{-j\pi f_0 t}}{2}) \\
&\quad * (\frac{e^{2j\pi f_0 t}}{2} + \frac{e^{-2j\pi f_0 t}}{2}).dt\ e^{j\omega t} \\
&= \int_{-T_0}^{T_0} (\frac{e^{(j2\pi f_0 + j\omega)t} + e^{(-j2\pi f_0 + j\omega)t}}{2} \\
&\quad + \frac{e^{(j3\pi f_0 + j\omega)t} + e^{(-j\pi f_0 + j\omega)t}}{4} \\
&\qquad + \frac{e^{(j\pi f_0 + j\omega)t} + e^{(-j3\pi f_0 + j\omega)t}}{4}).dt \\
&= \sin(\omega T_0)[\frac{2\omega}{-\omega_0^2 + \omega^2}] - \sin(\omega T_0)[\frac{\omega}{-(\frac{3}{2}\omega_0)^2 + \omega^2}] \\
&\quad - \sin(\omega T_0)[\frac{\omega}{-(\frac{\omega_0}{2})^2 + \omega^2}] \qquad \text{(A.2)}
\end{aligned}
$$

A.2. Visualization of Time and Frequency Domains

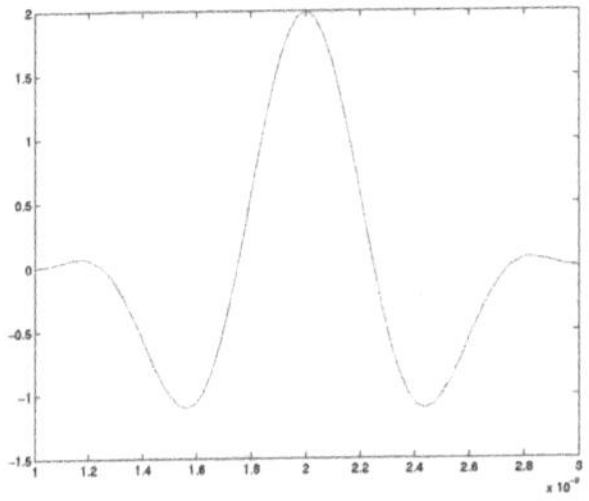

Figure A.1.: RC_2 in Time Domain.

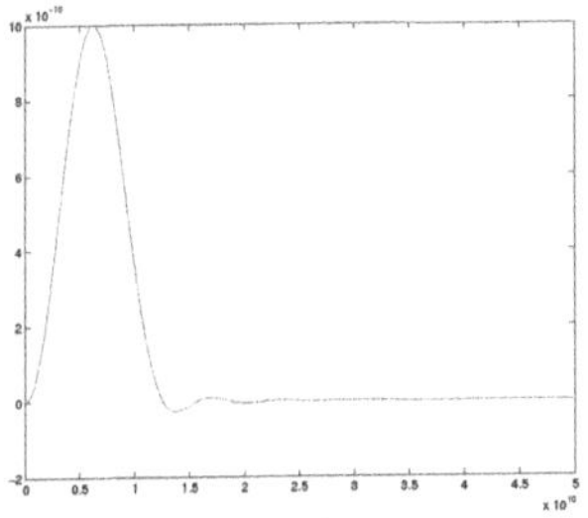

Figure A.2.: RC_2 in Frequency Domain.

B

Plane Waves

B.1. Electromagnetics

As far as we know, there are just four fundamental forces in nature: strong, electromagnetic, weak, and gravitational. The physical theory that describes electromagnetic forces is called electrodynamics. It was given its classical formulation by Maxwell over one hundred years ago with these forces being mediated by the photon.[1]

All electromagnetic phenomena are ultimately reducible to an elementary process of absorbing and emitting photons.[2] Such a process is treated in the field of quantum electrodynamics,[3] where the vacuum itself behaves like a dielectric.[4]

The dielectric constant[5] is also known as *relative permittivity*. This dimensionless parameter is measured (and set for various materials) in reference to its value in vacuum. It describes the material property which modifies the electrostatic force between the point charges in the medium. It is a scalar for isotropic medium and a second rank tensor for anisotropic medium.

[1] [Griffiths (2008), page 59-60]
[2] Ibid. page 61
[3] Ibid. page 65.
[4] Ibid. page 69.
[5] Which is a deprecated term in Engineering according to the *IEEE Standards Board (1997)*. https://ieeexplore.ieee.org/document/705931

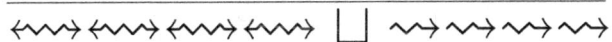

B.2. Electromagnetic Waves

This thesis solves the equations of the plane waves in the absence of free charges and currents. A basic feature of the Maxwell equations for the electromagnetic field is the existence of traveling wave solutions which represent the transport of energy from one point to another regardless of the *continued presence*[6] of the originating source whence this wave was set loose. The simplest and most fundamental electromagnetic waves are transverse, aka plane waves.[7] Such a model serves as an expression that states that the electromagnetic wave has the property of the electric and magnetic field components being perpendicular to one another and also perpendicular to the direction of wave propagation.[8]

The term *plane* indicates that the field vectors **E** and **H** at each point in space lie in a plane, with the planes at any two different points being parallel to each other.[9] Plane waves are good approximations to real waves in many practical situations. Radio waves at large distances from the transmitter, or from *diffracting*[10] objects, have negligible curvature and are well represented by plane waves. Much of optics utilizes the plane-wave approximation. More complicated electromagnetic wave patterns can be considered as a superposition of plane waves, so in this sense the plane waves are basic building blocks for all wave problems. Even when that approach is not followed, the basic ideas of propagation, reflection and refraction, help the understanding of other wave problems.[11]

Whether the wave has a high or low frequency, the same characteristics of propagation persists. For example, in *microwave engineering* the expression *plane wave* is present and also appears when microwaves propagate with a constant phase over a set of planes. In addition, if the magnitude also remains constant, the wave is called *uniform plane wave* with the properties of being transverse[12] and having stored energies equally

[6] It is important to understand the *source operation* in emitting electromagnetic waves as being separate from the *wave operation* in propagating independently and away from that same source which has produced it.

[7] [Jackson (1999), page 295]

[8] [Serway (1996), page 1000]

[9] [Clayton R. Paul (2000), page 348]

[10] Diffraction can be explained by wave theory. In *classical physics*, the diffraction phenomenon is described as the interference of waves according to the Huygens-Fresnel principle that treats each point in the wavefront as a collection of individual spherical wavelets. These characteristic behaviors (of media properties) are exhibited when a wave encounters an obstacle or a slit that is comparable in size to its wavelength. Similar effects occur when an electromagnetic wave travels through a medium with a refractive index, or when a sound wave travels through a medium with varying acoustic impedance. [Enders A. Robinson (2017), page 321]

[11] [Simon Ramo (1994), page 274]

[12] aka, transverse electromagnetic (TEM); *See Appendix B.3.*

divided between the **E** and **H** fields.[13] Therefore, for the purpose of understanding the contents of this thesis, consider the traveling waves as linearly polarized in a same analogous situation to that of a terminated transmission line in circuit theory. By analyzing the wave interaction at the interface between two media with field theory, it will be shown in the three chapters that we arrive at identical expressions for the reflection and transmission coefficients of waves and lines.[14]

As examples for microwave circuits and devices there are microwave transmission lines and waveguides. Microwave signals are propagated through these lines as electromagnetic waves and scattered from the associated junction of these lines to travel in well-defined direction or ports. The electromagnetic plane wave exists, however, in microwave transmission lines (but not in waveguides) and in the following applications: multi conductor lines, viz. coaxial lines, strip lines, microstrip lines, slot lines, and coplanar lines; where the mode of transmission is a TEM or quasi-TEM wave.[15]

B.3. Transverse Electromagnetic (TEM) Waves

In transmission lines, if the line conductors are not perfect conductors, the TEM mode cannot exist since there will be a longitudinal component of the electric field due to the line currents passing through the imperfect conductors. From a practical standpoint the line conductors, although not perfect conductors, will usually be sufficiently good conductors that this loss may be included as a reasonable approximation in the transmission formulation (which assumes a TEM field structure). Therefore, when having lossy line conductors while using the TEM-mode transmission-line formulation, it is assumed that the field structure is quasi-TEM (almost TEM). A lossy medium does not in itself preclude the existence of the TEM mode. In other words, a lossy medium and the TEM mode can co-exist so long as the medium is homogeneous.[16]

In an ideal coaxial line, an inner circular perfect conductor and an outer circular perfect conductor exist. The space between the two conductors is filled ideally with a uniform lossless homogeneous dielectric with two different potentials for the two conductors. The dominant mode of propagation in a symmetric coaxial line is TEM wave. And since TEM mode does not have a cutt-off frequency, a coaxial line is a broadband device.[17]

[13] [Annapurna Das (2008), page 35]
[14] [John D. Kraus (1999), page 189]
[15] [Annapurna Das (2008), page 51]
[16] [Clayton R. Paul (2000), page 478]
[17] [Annapurna Das (2008), page 51-2]

B.4. Scattering

B.4.1. Nuclear Physics

A monoenergetic[18] beam of particles can be represented by a plane wave since when a wave (of any type) hits a small obstacle, secondary waves (circular or spherical) are produced and move away from it into infinity. A plane wave, hence, undergoes scattering when it finds a region in which there exists a potential created by a nucleus.[19]

In problems of atomic and nuclear physics the detectors lie far away from the scattering centers compared to their dimensions, that is, they are in a region where the particles no longer feel the action of the potential. In that region, the asymptotic part of the scattered wavefunction is that which is dominant. And a detector which is placed in that region registers not only the presence of the plane wave, but also the particles scattered by the potential. Therefore, an outgoing spherical wave (which is created by the scattering center) will be added to the plane wave.[20]

The asymptotic form of a plane wave is represented (at large distances from the origin) through a reduced-form of spherical Bessel functions; with an outgoing and an incoming spherical wave components;[21] this is, however, only the radial aspect of the plane wave. Such a plane wave also has an angular aspect to it which is represented by an expansion in Legendre polynomials. This means that the plane wave can be understood as the sum of a set of partial waves, each one with orbital angular momentum and with a superposition of the two incoming and outgoing spherical components.[22] The asymptotic form of the wavefunction can be obtained if we observe that the presence of the potential has the effect of causing a *perturbation* in the outgoing part of the plane wave.[23]

Electromagnetic waves in free space are transversely polarized with degrees of freedom which play the role of the photon spin in *Nuclear Physics*. The transverse character of the electromagnetic field is related to the fact that quanta, photons, which appear after the field is quantized have zero mass. For massless particles there is no rest frame. Their free motion wavefunction is axially symmetric with respect to the axis of its wave vector, and, for nonzero spin, can be characterized by helicity, the spin projection onto the direction of motion. Helicity is Lorentz invariant for a massless particle. The two possible values of photon helicity, $\pm$, correspond to right (left) circular polarization. Their combination describe

[18] See also: *Monochromatic.*
[19] [Bertulani (2007), page 46]
[20] Ibid. page 47.
[21] Ibid. page 50.
[22] Ibid. page 49.
[23] Ibid. page 51.

two possible linear polarizations in the plane perpendicular to the axis of motion. The longitudinal polarization (which is along the wave vector) would have, however, zero helicity and is not allowed for the real massless photon.[24]

Quantization of the electromagnetic field in plane waves (while satisfying its own wave equation) is the recipe for transforming it into an arbitrary basis adjusted to any field geometry like spherical, cylindrical waves or waves in a cavity of arbitrary shape.[25]

B.4.2. Classical Physics

In Classical Physics, scattering is studied using the *Excitation of Media*[26].

[24] Ibid. page 219-20.
[25] Ibid. page 224.
[26] See Appendix F.

C

Boundary Conditions

C.1. Constitutive Relations

To allow a unique determination of the field vectors, Maxwell equations were supplemented by relations which describe the behavior of the medium which is under the influence of the field. These subsidiary relations are called *constitutive relations* which can be established by experimentation or deduced from atomic theory. If the field vectors are linearly related so that the principle of superposition applies, the medium is said to be *linear*. A combination of Maxwell equations and *linear constitutive relations* forms the basis of *linear electrodynamics*, which is the main concern of this thesis. The linear relations among the field vectors may be either algebraic or involve differentiation and integration, depending upon the medium under consideration. In general, they are rather difficult to formulate in the time domain,[1] therefore, one finds first (by mathematical modeling, aka formulation) the solution in the frequency domain and then uses a numerical method and/or tool[2] to calculate (and visualize) the values back in time domain.

C.2. Homogeneity

If the dielectric constant ε, the relative permeability μ, and the conductivity σ are scalars (or if they simply have the same values at every point in space), the medium is said to be *homogeneous*. On the other hand, if the electrical and magnetic properties of a medium depend upon the

[1] [Chen (1983), page 61]
[2] As I have used the FFT tool in Matlab.

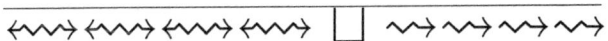

directions of field vectors, the medium is called *anisotropic*.[3] However, across any surface which bounds one medium from another the constitutive parameters such as ε, μ, or σ may change abruptly, and we may expect corresponding changes in the field vectors. In order to continue the solution of Maxwell equations from one region to another so that the resulting solution is unique and valid everywhere, we need *boundary conditions* to impose on the field vectors at the interface.[4]

[3] Ibid. page 62.
[4] Ibid. page 63.

D

Monochromatic Fields

Since the Maxwell equations together with the linear constitutive relations form a linear system of partial differential equations, no generality is lost by considering the sinusoidal time-varying fields alone. In fact, according to Fourier analysis, an arbitrary time-varying function can always be represented as a superposition of sinusoidal functions. If a field vector varies sinusoidally in time with a single angular frequency ω, the field is said to be *monochromatic.*[1]

[1] Ibid. page 67.

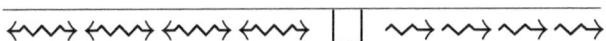

E

Stationary Media

In this thesis, the solutions for Maxwell equations were obtained only for stationary media. If I, otherwise, were to have a moving media (i.e. moving plate) in my application, then Einstein's special theory of relativity has to be taken into consideration. In this case, Maxwell equations must be made *covariant* under the Lorentz Transformation[1]; the latter could be established from the principle of the constancy of the velocity of light.[2] After that the derivation of the transformation formulas for sources and field vectors in two systems moving with a relative uniform velocity must be delivered.[3] And finally, Maxwell equations must be demonstrated to be Lorentz covariant before proceeding into deriving the specific equations of my own special application.[4]

[1] Ibid. page 74.
[2] Ibid. page 78.
[3] Ibid. page 81.
[4] This, however, is not part of this thesis.

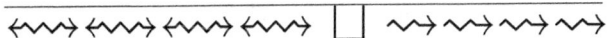

F

Excitation of Media

In this thesis, the propagation of electromagnetic waves in various media as well as the reflection and transmission of these waves from the surface of such media were treated without discussing how these media were excited. One possible method of approach to consider media excitation is by using the *dyadic Green function*, which yields the field directly in terms of the source current.[1]

Excitation (aka Coupling) takes place when electromagnetic fields or current sources are generated by a given distribution of sources or electromagnetic fields, respectively. We can simply consider the source current as an input and the electric field as an output or vice versa.[2] The simplest radiating source for example is an oscillating electric dipole.[3]

A receiving antenna, for example, may be viewed as any metal object that scatters an incident electromagnetic field. As a result of scattering, an electric current appears on the antenna's surface. The current in turn creates a corresponding electric field.[4]

In a nonrelativistic system of charged particles that interact with electromagnetic fields, the capability of absorbing and emitting photons arises. The electromagnetic interaction is, however, relatively weak compared to the strong forces and the relevant parameter is the fine structure constant.[5] The physical properties of the system become apparent in specific features of the current which is interacting with the radiation field. In contrast to this, the electromagnetic field is expressed universally. The

[1] [Chen (1983), page 340]
[2] Ibid. page 342.
[3] Ibid. page 349.
[4] [Makarov (2002), chapter 2]
[5] [Bertulani (2007), page 224-5]

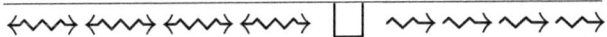

quantized form of the *vector potential* contains an exponent which plays the role of "orbital" wavefunction of the photon, whereas the polarization vectors characterize the intrinsic state of the photon, that is, its spin. Thus, the photon is an object with momentum and spin. This is how we can now expand the plane wave into partial waves with a certain orbital momentum. Each partial wave can be combined with the spin vector into a state with the total angular momentum of the photon.[6]

For example, the calculation of the decay rate for gamma emission requires the use of a quantum theory for the radiation. But it is instructive to present the correct quantum results as an extension of calculations based on classical electrodynamics. Classically, the electromagnetic radiation emitted by a system is the result of the variation in time of the charge density or of the distribution of charge currents in the system. The energy is emitted in two types of multipole radiation: the electric and the magnetic. Each one of them is expressed as a function of the corresponding multipole moments, being the quantities that contain the variables (charge and current) of the system. As the speed of the charges inside the nucleus is much smaller than the speed of light, the magnetic multipole moments are in general much smaller than the corresponding electric moments of the same order, except when selection rules forbid the existence of decay through the corresponding electric multipole. In quantum mechanics, however, the energy is emitted not continually but in packets of energy and that is how the decaying nucleus should be treated.[7]

In *nuclear reaction*, the story is different because the occurrence of a nuclear reaction proceeding through a given reaction channel[8] leads to a modification of the outgoing part of the wave. This case does not count as an *elastic* scattering (aka elastic channel) but rather as an, *inelastic* channel.[9] The perturbation potential is that which causes the "transition" from the entrance to the exit channel. This transition can be understood as an additional interaction to the average behavior of the potential, and, in this sense, can be written as the difference between the total potential in the exit channel and the potential of the optical model[10] in that same channel. The simplest treatment that we can do for a direct reaction is to consider the incident beam as a plane wave whose only interaction with the target is through the perturbation potential that causes the reaction. The emerging beam is also treated as a plane wave.[11]

---

[6] Ibid. page 235-6
[7] Ibid. page 238-9
[8] Each of the branches of the reaction which can occur is referred to as a *channel*. [Ibid. page 259]
[9] Ibid. page 267
[10] Because of the analogies to optical phenomenon, it may be no surprise that the potentials used in nuclear scattering are called "optical potentials" and that one speaks of "optical models". [Ibid. page 276]
[11] Ibid. page 302

G

Advanced Calculus

G.1. Vectors and Matrices

If we take into consideration the higher-level aspects of calculus, we will arrive at the central idea of the differential calculus which is the approximation of a nonlinear function by a linear one. Geometrically, one is approximating a curve or surface by a tangent line or plane. Such method reduces the problems of calculus to ones of algebra which are associated with lines and planes, i.e. linear algebra.[1]

A vector in space has a magnitude (length) and a direction but no fixed location. We can thus represent any vector by many other directed line segments in space, all having the same length and direction. All coordinate systems have *unit vectors* which are vectors of length 1 and have the directions of the coordinate axes.[2]

Vectors in space are an *ordered triple of numbers* which can also be represented as matrices. A *zero vector*, however, is a vector which have zero length and is orthogonal and parallel to all other vectors.[3] And the angles which the unit vectors have in reference to any other specific vector are called *direction angles* of that vector; each is accompanied by a *direction cosine*.[4]

The *dot product* is known as the *inner product* whereas the *vector product* as the *cross product*. The latter is usually specified by a right-hand

[1] [Kaplan (2003), page 1]
[2] Ibid. page 2.
[3] Ibid. page 3.
[4] Ibid. page 4.

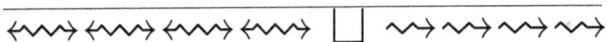

rule where the *ordered triple of vectors*[5] are called a *positive* triple if the vectors can be moved continuously to attain the respective directions of their unit vectors eventually without making one of the vectors lie in a plane parallel to the other two. The triple is called *negative* if their order is changed in such a way that the first and second vectors are now associated with the second and first unit vectors respectively.[6]

Two vectors in space are said to be *linearly independent* if they cannot be represented by directed line segments on the same line. Otherwise, they are said to be *linearly dependent* or *collinear*. When one of them is a zero vector, the vectors are considered to be linearly dependent. Also, if the cross product between two vectors equals to a zero vector, then both vectors are linearly dependent. If we extend these definitions to three vectors in space, we will have linearly independent vectors when they all cannot be represented by directed line segments in the same plane. Otherwise, they are said to be linearly dependent or *coplanar*.[7]

In linear algebra, if we have in some equation a summation of two *mathematical unit vectors* which equals to a zero vector, then both vectors are linearly dependent[8]. The same goes for a combination of three mathematical unit vectors. In other words, in two-dimensional space, any mathematical combination of three unit vectors (or above) must be linearly dependent and in three-dimensional space, any mathematical combination of four vectors (or above) is also linearly dependent. So the *normal vector* of any plane[9] is a nonzero vector which is perpendicular to the plane.[10]

Every linear equation represents a plane and has the following characteristics:.[11]

$$A(x-x_1)+B(y-y_1)+C(z-z_1) = 0 \tag{G.1}$$

$$Ax+By+Cz+D = 0 \tag{G.2}$$

where

$$\mathbf{n} = A\underline{i}+B\underline{j}+C\underline{k} \tag{G.3}$$

[5] Which is not the same as an ordered triple of numbers; the latter is one single vector with three (or two) components.

[6] [Kaplan (2003), page 5]

[7] Ibid. page 6.

[8] i.e. not real unit vectors; not physical unit vectors.

[9] This is obviously in three-dimensional space whether we have one, two or three vectors because the normal vector itself extends the plane out of its two-dimensional realm into three spatial dimensions. In two-dimensional space we have a normal vector which extends the one-dimensional vector into two spatial dimensions (a plane).

[10] [Kaplan (2003), page 8]

[11] Ibid. page 9.

is the normal vector on that plane. The visual representation thereof is demonstrated in Figure 1 below.

With (x_1,y_1,z_1) as the endpoint of vector $\vec{R}'$ and hence represents its ordered triple of numbers.[12] Also notice that each variation of x, y and z in (x,y,z) generates a 3-tuple which connects to (x_1,y_1,z_1) through the vectors: $\vec{V} = \vec{R} - \vec{R}'$ and $-\vec{V} = \vec{R}' - \vec{R}$. And whether these two vectors are linearly dependent or not, the mere existence of the same normal vector for both indicates that they lie in one single plane. By decoupling the normal vector from equation (G.1) above, we get:[13]

$$(x - x_1)\mathbf{e}_x + (y - y_1)\mathbf{e}_y + (z - z_1)\mathbf{e}_z = \vec{V} \qquad \text{(G.4)}$$

The normal vector, however, is perpendicular to $\vec{V}$ and $-\vec{V}$ but not, generally speaking, perpendicular to $\vec{R}$ and/or $\vec{R}'$.

G.2. Determinants

Any three linearly independent vectors in space have the elements of a determinant which has a value equal $\pm$ the volume of a parallelepiped[14] whose edges represent these three vectors. The negative volume indicates the negatively ordered triple of vectors inside the determinant and at the edges of the parallelepiped.[15]

These three vectors are, hence, linearly independent if the determinant has a volume and does not equal to zero. And for two vectors the cross product delivers the $\pm$ area of a parallelogram; where the vectors are linearly independent. The triple (or double) of vectors is positive when crossing two of its vectors with an angle ϕ that lies between 0 and π and negative when ϕ is between π and 2π.[16] This all takes place mathematically speaking when interchanging two rows (or columns) in a determinant. By doing so, the value of the determinant is multiplied by (-1). We call this a *permutation* of the rows (or columns) which we term *even* or *odd* according to the number of interchanges; the former reverses the sign twice and leaves the determinant unchanged.[17]

[12] Notice here that I use a third different vector notation in addition to the conventions which I have followed throughout the pages of this book; see footnotes 1 and 2 in chapter 1. I do so to simply highlight the mathematical context of these equations regardless of the time and/or frequency domains.

[13] Note that $\mathbf{e}_x = \underline{i}$, $\mathbf{e}_y = \underline{j}$, $\mathbf{e}_z = \underline{k}$.

[14] A parallelepiped is a solid body of which each face is a parallelogram.

[15] [Kaplan (2003), page 11]

[16] Ibid. page 12.

[17] Ibid. page 11.

G.3. Linear Equations

A system of n equations with n unknowns (or less) can be solved using Cramer's Rule which asserts that when the determinant $\neq 0$, then the following equations

$$a_{11}x_1 + a_{12}x_2 + \cdots + a_{1n}x_n = k_1,$$

$$a_{n1}x_1 + a_{n2}x_2 + \cdots + a_{nn}x_n = k_n, \tag{G.5}$$

have the unique solution

$$x_1 = D_1/D, \cdots, x_n = D_n/D \tag{G.6}$$

where

$$D_1 = \tag{G.7}$$

and

$$D = \tag{G.8}$$

is the determinant of matrix A=(a_{ij}) which is called the *coefficient matrix* of the set of equations.[18]

G.4. Inverse of a Square Matrix

(1) B is an inverse of A if $AB = I$, where $B = A^{-1} = 1/A$. (2) A matrix has at most one inverse. (3) If A has an inverse, then $det(A) \neq 0$. This means that $det(A)det(B) = det(I) = 1$. Conversely, if $det(A) \neq 0$, then A has an inverse. (4) A matrix A having an inverse is said to be *nonsingular*. Otherwise, it is said to be *singular*. (5) If $k < n$, A is an $n \times k$ matrix, and B is a $k \times n$ matrix, then AB is an $n \times n$ matrix and AB is singular.[19]

G.5. Eigenvalues of a Square Matrix

We have $A\vec{v} = \lambda\vec{v}$ for some scalar λ. If this occurs, we say that λ is an *eigenvalue* of A and that $\vec{v}$ is an *eigenvector* of A. The concept of eigenvalue has important applications in many branches of Physics.[20] The *spectrum* of light in an atom (in a nucleus) has frequencies that correspond to the eigenvalues of a matrix. However, in our treatise, the eigenvalues are important in the solution of linear differential equations.

[18] Ibid. page 13-5.
[19] Ibid. page 26-9.
[20] As in our specific application in this book

So we have $A\vec{v} = \lambda I\vec{v}$ in the form $(A - \lambda I)\vec{v} = \vec{0}$ where $\vec{v}$ is an eigenvector of A when $v_1, \cdots, v_2$ form a nontrivial[21] solution of the set of homogeneous linear equations which have the following determinant

$$= 0 \tag{G.9}$$

When expanded, determinant $G.9$ becomes an algebraic equation of degree n for λ, called the *characteristic equation* of the matrix A. We call B a *diagonal* matrix and write $B = diag(\lambda_1, \cdots, \lambda_n)$. Then B has the characteristic equation

$$= 0 \tag{G.10}$$

or

$$(\lambda_1 - \lambda)(\lambda_2 - \lambda)\cdots(\lambda_n - \lambda) = 0 \tag{G.11}$$

where $\lambda_1, \cdots, \lambda_n$ are the eigenvalues of B. We say B is similar to A if

$$B = C^{-1}AC \tag{G.12}$$

and A is similar to B as a consequence thereof. Hence we speak of similar matrices A, B. When A and B are similar, then A and B have the same characteristic equation because

$$det(B - \lambda I) = det(A - \lambda I)$$

with A and B having the same eigenvalues as well. Here we can view C as the matrix containing several eigenvectors with n distinct real eigenvalues $\lambda_1, \cdots, \lambda_n$ associated with the eigenvector $\vec{v}_1, \cdots, \vec{v}_n$. Therefore.[22]

$$AC$$

G.6. The Transpose

The transpose of a matrix is obtained by interchanging rows and columns. A matrix which equals to its transpose is called a *symmetric* matrix. It is also important to know that every diagonal matrix is symmetric. Symmetric matrices are useful in discussing quadratic forms, that is, algebraic expressions of the form

$$Q(\vec{x}) = \tag{G.13}$$

And with such an algebraic expression, we can achieve symmetry for the coefficient matrix (a_{ij}) through splitting the combined term into two equal numbers. So instead of having $a_{12}x_1x_2$ and $a_{21}x_2x_1$, we have

[21] A trivial solution is when $v_1 = v_2 = \cdots = v_n = 0$.
[22] Ibid. page 35-7.

$\frac{a_{12}+a_{21}}{2}x_1x_2$ and $\frac{a_{12}+a_{21}}{2}x_2x_1$. For each assignment of numerical values to $x_1, \cdots, x_n$, equation $G.13$ has a numerical value Q.

We can generate such numbers for each vector $\vec{x}$ by considering x_i as its substituted numerical values and x_j as the numerical values for its transpose, i.e. row vector in case $\vec{x}$ were a column vector.[23]

G.7. Orthogonal Matrices

A matrix is orthogonal if $AA' = I$. Therefore, A is orthogonal if and only if $A^{-1} = A'$, that is, if and only if the inverse of A equals the transpose of A. Thus every orthogonal matrix is nonsingular. This makes the row vectors, or the column vectors, of the matrix as mutually perpendicular unit vectors. To phrase this algebraically:

$$a_{i1}^2 + \cdots + a_{in}^2 = 1, \qquad i = 1, \cdots, n \tag{G.14}$$

$$a_{i1}a_{j1} + \cdots + a_{in}a_{jn} = 0, \qquad i \neq j, \; i = 1, \cdots, n, \; j = 1, \cdots, n \tag{G.15}$$

$$a_{1j}^2 + \cdots + a_{nj}^2 = 1, \qquad j = 1, \cdots, n \tag{G.16}$$

$$a_{1j}a_{1k} + \cdots + a_{nj}a_{nk} = 0, \qquad j \neq k, \; j = 1, \cdots, n, \; k = 1, \cdots, n \tag{G.17}$$

Equation $G.14$ above states that the row vectors are unit vectors and equation $G.15$ states that different row vectors are orthogonal while equations $G.16$ and $G.17$ are for the column vectors.[24]

Orthogonal matrices are important in studying changes of coordinates and when they are placed as a matrix of eigenvectors A between two similar matrices C and B, one speaks of *orthogonally congruent* matrices (i.e. A and B).

$$B = A^{-1}CA$$

And if C is symmetric, then C is orthogonally congruent to a diagonal matrix B. This previous special case reduces equation $G.13$ into

$$Q(\underline{x}) = \lambda_1 y_1^2 + \cdots + \lambda_n y_n^2 \tag{G.18}$$

[23] Ibid. page 39-40.
[24] Ibid. page 40-3.

with $\lambda_1, \cdots, \lambda_n$ as eigenvalues of B (and C). Here, it is not even necessary to find the matrix A. We can interpret $x_1, \cdots, x_n$ as Cartesian coordinates in n-dimensional space. Then $y_1, \cdots, y_n$ are simply new Cartesian coordinates in n-dimensional space with the same origin. For $n = 2$, $Q(\underline{x}) = 1$ and hence represents a conic section of an ellipse or hyperbola. In fact, in the language xxxxxxx that every second-degree equation $Ax^2 + By^2 = 1$ by an appropriate rotation of axes in the xy-plane.[25]

G.8. Analytic Geometry

In *Analytic Geometry* there is no direct physical meaning for the entity but rather a purely mathematically formulated ones. These entities can then be interpreted into the physical realm. In a *Euclidean n-dimensional space*, E^n is defined as a space having n coordinates $x_1, \cdots, x_n$ with n being a fixed positive integer. A point P of the space is defined by an ordered $n-tuple$ $(x_1, \cdots, x_n)$. Point $(0, \cdots, 0)$ is the origin O. A vector $\underline{v}$ in n-dimensional space is defined to be an ordered n-tuple $(v_1, \cdots, v_n)$. So, both points and vectors are represented by n-tuples with each point we can associate the vector $\vec{OP} = (x_1, \cdots, x_n) = \underline{x}$. Conversely, to each vector $\vec{x}$, we can assign the point P and then consider it springing out from the origin $\underline{x} = \vec{OP}$. Vectors can also be interpreted as matrices either as row vectors or as column vectors.[26]

We denote the set of all vectors by V^n. These vectors are linearly independent if an equation $c_1\underline{v}_1 + \cdots + c_k\underline{v}_k = \underline{0}$ can hold only if $c_1 = \cdots = c_k = 0$. Otherwise, the vectors are said to be linearly dependent.[27]

$$\begin{bmatrix} a_{11} & \cdots & a_{1k} \\ \vdots & \ddots & \vdots \\ a_{n1} & \cdots & a_{nk} \end{bmatrix} \begin{bmatrix} c_1 \\ \vdots \\ c_k \end{bmatrix} = c_1\underline{v}_1 + \cdots + c_k\underline{v}_k \tag{G.19}$$

Accordingly the vectors are linearly independent if and only if A is non-singular, that is, when $detA \neq 0$.

We have the *Cauchy-Schwarz inequality*

$$|\underline{u}.\underline{v}| \leq |\underline{u}||\underline{v}| \tag{G.20}$$

as well as the *triangle inequality*

$$|\underline{u} + \underline{v}| \leq |\underline{u}| + |\underline{v}| \tag{G.21}$$

[25] Ibid.
[26] Ibid. page 46.
[27] Ibid. page 47-50.

where the equality holds in both equations only if the two vectors $\underline{u}$ and $\underline{v}$ are linearly dependent because

$$\cos\theta = \frac{\underline{u}.\underline{v}}{|\underline{u}||\underline{v}|}, \qquad 0 \leq \theta \leq \pi \tag{G.22}$$

Therefore for three points in E^n we have

$$|\overrightarrow{P_1P_3}| \leq |\overrightarrow{P_1P_2}| + |\overrightarrow{P_2P_3}| \tag{G.23}$$

An orthogonal system of unit vectors is called an orthonormal system. It is noteworthy to mention that it is always possible to construct an orthogonal system of vectors through a process which is called: *Gram-Schmidt* orthogonalization process.

G.9. Applications in Physics

In reality, V^n is the only Euclidean n-dimensional vector space. Such a generalization of vectors to n-dimensional space instead of three is valuable even in Physics and not only in Mathematics. We see this in the equations of mechanical systems with *N degrees of freedom*; and in the kinetic theory of gases with an n-dimensional *phase space* (here n may be as large as 10^{23}); and in relativity with a 4-dimensional spacetime is needed; and in quantum mechanics where the vector space of infinite dimensions is closely related to the theory of *Fourier Series.*[28]

G.10. Linear Mapping

A function is called a *mapping*. Furthermore, if $\underline{x}_1$ and $\underline{x}_2$ are in V^n and c_1, c_2 are scalars, then

$$A(c_1\underline{x}_1 + c_2\underline{x}_2) = c_1(A\underline{x}_1) + c_2(A\underline{x}_2) \tag{G.24}$$

has a mapping which is

$$A\underline{x} = y \tag{G.25}$$

This mapping assigns to each linear combination $c_1\underline{x}_1 + c_2\underline{x}_2$ the corresponding linear combination of the values assigned to $\underline{x}_1$ adn $\underline{x}_2$. We call such a mapping: Linear. Hence, the linear mapping T from V^n into V^m is done using $T(\underline{x}) = A\underline{x}$. The set of all $\underline{x}$ for which $T(\underline{x}) = \underline{0}$ is called the *kernel* of T. The identity Matrix is called the *identity mapping* and is clearly linear and one-to-one. This is contrasted by the *zero mapping* which is linear but not one-to-one.[29]

[28] Ibid. page 52-3.
[29] Ibid. page 55-8.

G.11. Rank of Matrix

A subspace of V^n is a collection W of vectors of V^n. If these vectors are linearly independent then they form the basis of W. Otherwise, the basis can be determined by the *Gaussian elimination* method. The *rank of a matrix* A is the maximal number of linearly independent row (or column) vectors of A.[30]

[30] Ibid. page 62-3.

References

Bibliography

Annapurna Das, S. K. D. (2008). *Microwave Engineering*. McGraw-Hill.

Bertulani, C. A. (2007). *Nuclear Physics in a Nutshell*. Princeton University Press.

Chen, H. C. (1983). *Theory of Electromagnetic Waves*. McGraw-Hill.

Clayton R. Paul, Keith W. Whites, S. A. N. (2000). *Introduction to Electromagnetic Fields*. McGraw-Hill.

Enders A. Robinson, D. C. (2017). *Basic Geophysics*. Society of Exploration Geophysicists.

Griffiths, D. (2008). *Introduction to Elementary Particles*. Wiley-VCH.

Jackson, J. D. (1999). *Classical Electrodynamics*. John Wiley and Sons.

John D. Kraus, D. A. F. (1999). *Electromagnetics with Applications*. McGraw-Hill.

Kaplan, W. (2003). *Advanced Calculus*. Addison-Wesley.

Makarov, S. (2002). *Antenna and EM Modeling with MATLAB*. Wiley-Interscience.

Serway, R. A. (1996). *Physics for Scientists and Engineers*. Saunders College Publishing.

Simon Ramo, John R. Whinnery, T. v. D. (1994). *Fields and Waves in Communication Electronics*. John Wiley and Sons.

Index

www.ingramcontent.com/pod-product-compliance
Lightning Source LLC
Chambersburg PA
CBHW071256220526
45468CB00001B/153
9781791530709